Mobile Cranes

By James Headley

A safety handbook for operators, riggers, supervisors
and other personnel who use mobile cranes
to accomplish their work.

Crane Institute
of America, Inc.

The Standard in Crane Training & Operator Certification

Mobile Cranes

by James Headley

A publication of:
Crane Institute of America Publishing & Products, Inc.,
3880 St. Johns Parkway
Sanford, Florida 32771
email: info@craneinstitute.com
web: www.craneinstitute.com

Fifth Edition. Copyright © 2003, 2002, 2001, 2000, 1999 Crane Institute of America Publishing & Products, Inc.

ISBN: 0-9744279-0-X

Crane Institute of America, Inc. has made strenuous efforts to ensure the reliability of the information provided in this publication. However, the publisher or author can accept no liability for inaccuracies or incompleteness. This publication should not be considered a substitute for relevant regulations and standards or manufacturers' specific recommendations.

The graphics used to illustrate information are not intended to be exact representations, but rather they are added to help the reader's understanding.

Load charts are used for explaining principles and are not to be used for actual lifting or other field use.

For information on other publications and products, or for details on Crane Institute's nationwide training and certification programs, call 1-800-832-2726. Outside USA call (407) 322-6800.

Table of Contents

Preface

by Howard Shapiro, P.E.

There is a lot in this handbook, a lot more than you would expect from its small size. From beginning to end, clear graphic images have been used in place of complicated explanations – the high quality drawings make this a standout among handbooks on mobile cranes.

Chapter One presents names and illustrations for all the significant types, configurations, and components of mobile cranes. You can't expect to study a complicated subject without first clarifying the terms that must be used, and this is a great start. The idea that rating charts, the subject of Chapter Two, are not that complicated has led many people to make unfortunate errors. In order to know how to use rating charts, it is necessary to understand many details and to have knowledge about crane operational concepts; these things certainly are complicated. The graphic presentation of these details and concepts break down that complexity into common sense explanations. This chapter should become the crane operator's Bible – they should study it over and over again, and keep it close at hand.

Chapter Three deals with jobsite conditions that reduce a crane's capacity and that can bring grief. Some are conditions that need to be anticipated and taken into account, while others are situations that should be avoided. The next chapter illustrates the items that must be covered by pre-operational inspections. Following this are chapters about crane setup, working near power lines, operating practices and procedures, and hoisting personnel. The book ends with information tables for common mobile crane accessories, wire rope, slings and other hardware.

This handbook is a must for crane operators and their supervisors, and an important source of knowledge for riggers and anyone else who depends on mobile cranes. Although it contains mostly graphics and few words, readers will not find this a lightweight presentation. It thoroughly covers most of the complex issues of crane technology needed at the jobsite for safe and effective use of these versatile lifting machines.

Contents

Telescoping Boom Crane Types
Industrial Cranes

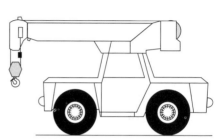

Carrydeck Crane
(Rotating Boom)

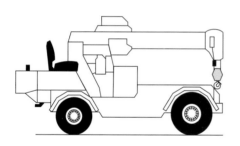

Carrydeck Crane
(Fixed Boom)

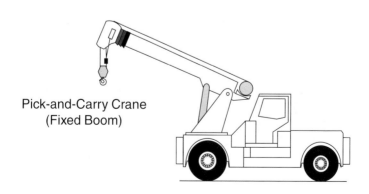

Pick-and-Carry Crane
(Fixed Boom)

Telescoping Boom Crane Types
Rough Terrain Cranes

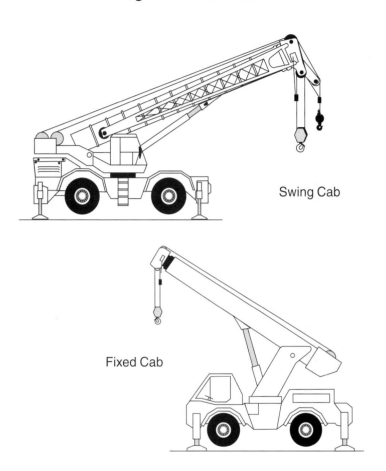

Swing Cab

Fixed Cab

All Terrain Crane

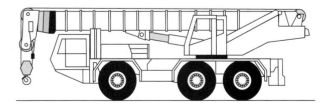

Telescoping Boom Crane Types
Commercial Truck-Mounted Cranes

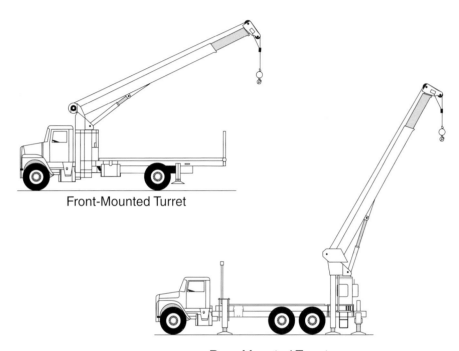

Front-Mounted Turret

Rear-Mounted Turret

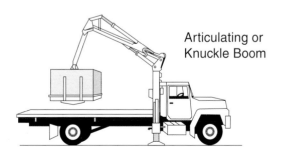

Articulating or
Knuckle Boom

Telescoping Boom Crane Types
Commercial Truck-Mounted Cranes

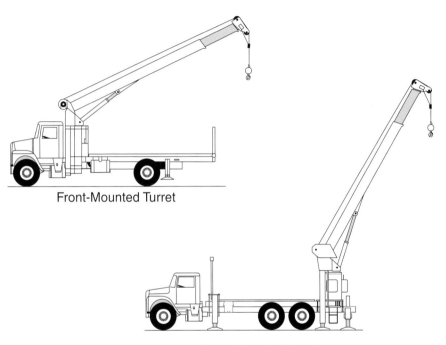

Front-Mounted Turret

Rear-Mounted Turret

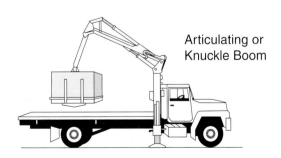

Articulating or
Knuckle Boom

Telescoping Boom Crane Types
Boom Attachments

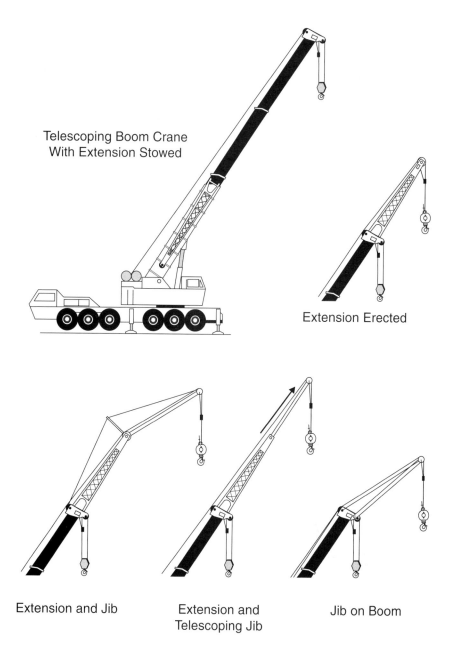

Telescoping Boom Crane
With Extension Stowed

Extension Erected

Extension and Jib

Extension and
Telescoping Jib

Jib on Boom

Telescoping Boom Crane Types
Boom Attachments

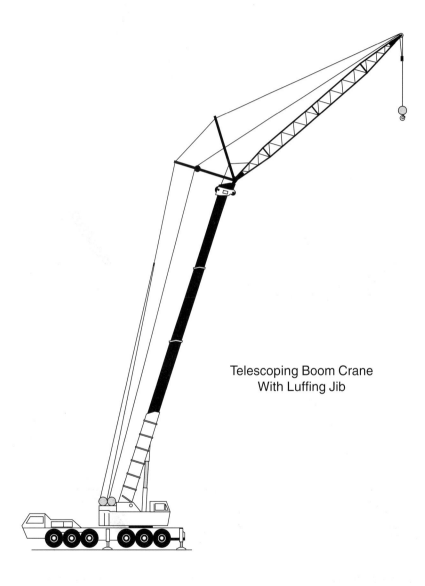

Telescoping Boom Crane
With Luffing Jib

Telescoping Boom Crane Components
Rough Terrain Cranes

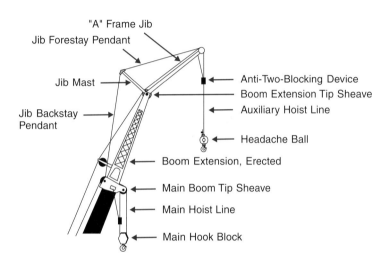

"A" Frame Jib
Jib Forestay Pendant

Jib Mast →

Anti-Two-Blocking Device
Boom Extension Tip Sheave

Jib Backstay
Pendant →

Auxiliary Hoist Line

Headache Ball

Boom Extension, Erected

Main Boom Tip Sheave

Main Hoist Line

Main Hook Block

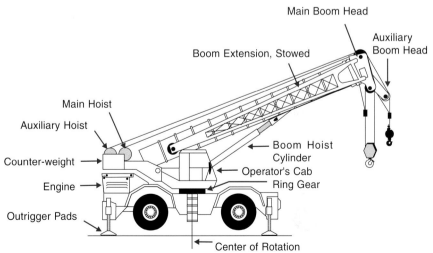

Main Boom Head

Auxiliary
Boom Head

Boom Extension, Stowed

Main Hoist

Auxiliary Hoist

Counter-weight →

Boom Hoist
Cylinder

Engine →

Operator's Cab
Ring Gear

Outrigger Pads →

Center of Rotation

Telescoping Boom Crane Components
Carrier Mounted Cranes

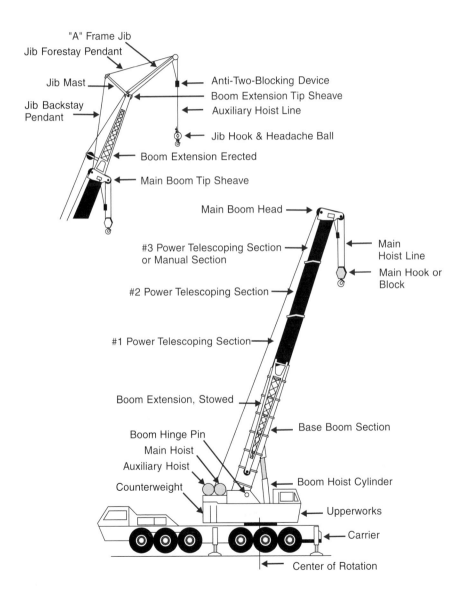

"A" Frame Jib
Jib Forestay Pendant

Jib Mast

Jib Backstay
Pendant

Anti-Two-Blocking Device
Boom Extension Tip Sheave
Auxiliary Hoist Line

Jib Hook & Headache Ball

Boom Extension Erected

Main Boom Tip Sheave

Main Boom Head

#3 Power Telescoping Section
or Manual Section

Main
Hoist Line

Main Hook or
Block

#2 Power Telescoping Section

#1 Power Telescoping Section

Boom Extension, Stowed

Base Boom Section

Boom Hinge Pin
Main Hoist
Auxiliary Hoist
Counterweight

Boom Hoist Cylinder

Upperworks

Carrier

Center of Rotation

9

Telescoping Boom Crane Components
Boom Truck Cranes

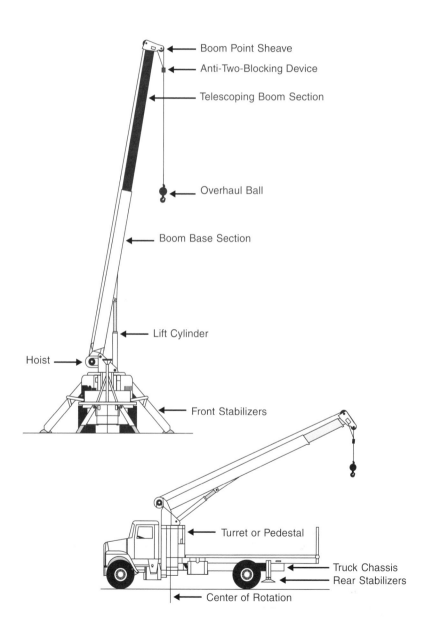

Boom Point Sheave

Anti-Two-Blocking Device

Telescoping Boom Section

Overhaul Ball

Boom Base Section

Lift Cylinder

Hoist

Front Stabilizers

Turret or Pedestal

Truck Chassis

Rear Stabilizers

Center of Rotation

Lattice Boom Crane Types
Crawler and Carrier Mounted Cranes

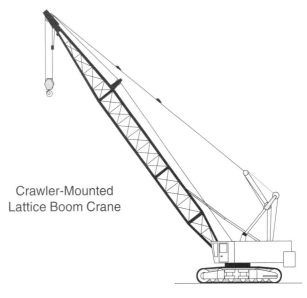

Crawler-Mounted
Lattice Boom Crane

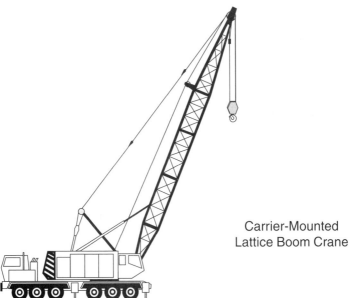

Carrier-Mounted
Lattice Boom Crane

Lattice Boom Crane Types
Heavy Lift Attachments

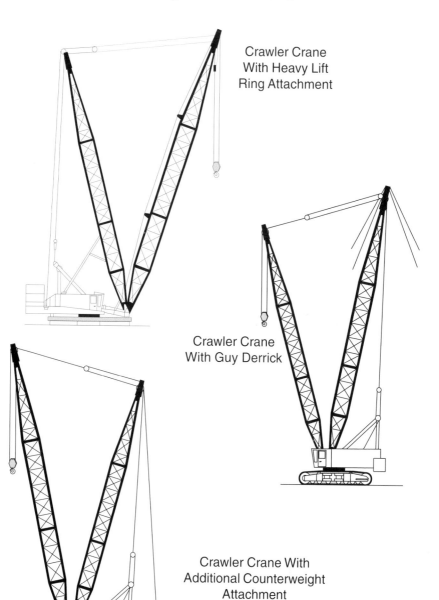

Crawler Crane
With Heavy Lift
Ring Attachment

Crawler Crane
With Guy Derrick

Crawler Crane With
Additional Counterweight
Attachment

Lattice Boom Crane Types
Boom Attachments

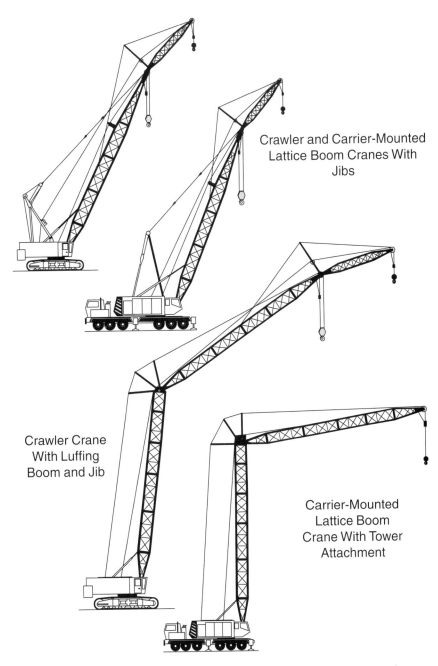

Crawler and Carrier-Mounted
Lattice Boom Cranes With
Jibs

Crawler Crane
With Luffing
Boom and Jib

Carrier-Mounted
Lattice Boom
Crane With Tower
Attachment

Lattice Boom Crane Components
Carrier Mounted Cranes

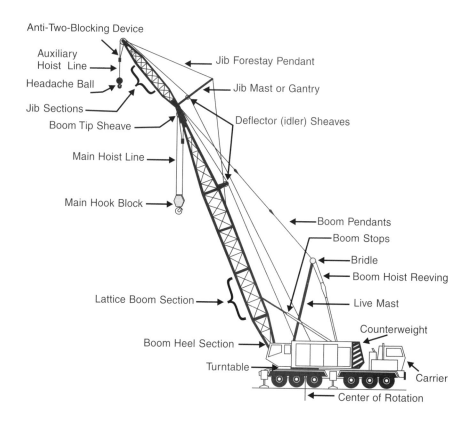

Anti-Two-Blocking Device

Auxiliary Hoist Line

Headache Ball

Jib Sections

Boom Tip Sheave

Main Hoist Line

Main Hook Block

Jib Forestay Pendant

Jib Mast or Gantry

Deflector (idler) Sheaves

Boom Pendants

Boom Stops

Bridle

Boom Hoist Reeving

Lattice Boom Section

Live Mast

Counterweight

Boom Heel Section

Turntable

Carrier

Center of Rotation

Lattice Boom Crane Components
Crawler Mounted Cranes

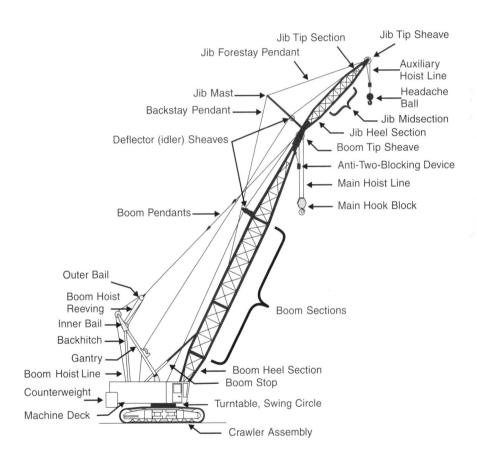

15

Contents

Contents

Crane's Center of Gravity

 Center of gravity is the point in an object where its weight may be considered to be concentrated. Like all objects, major crane components have a center of gravity.

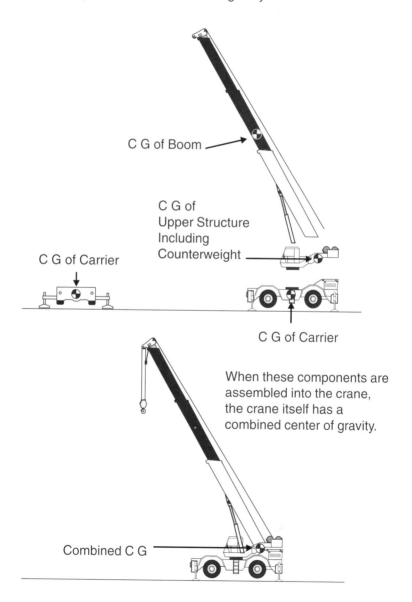

C G of Boom

C G of Upper Structure Including Counterweight

C G of Carrier

C G of Carrier

When these components are assembled into the crane, the crane itself has a combined center of gravity.

Combined C G

Principle of Leverage

The principle of leverage is used to determine rated capacities.

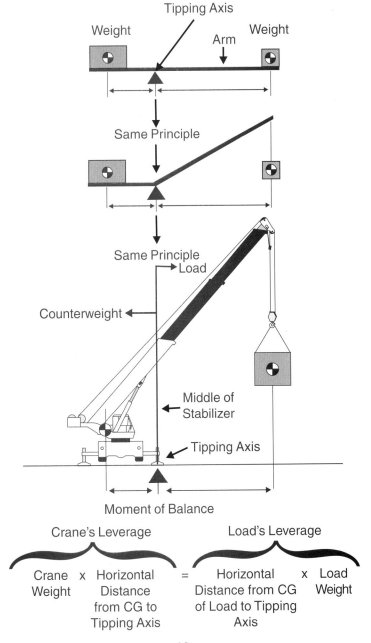

Crane's Leverage				Load's Leverage	
Crane Weight	x	Horizontal Distance from CG to Tipping Axis	=	Horizontal Distance from CG of Load to Tipping Axis	x Load Weight

Leverage and Stability

As the upper structure rotates, the crane's center of gravity moves closer to its tipping axis. The movement of the crane's center of gravity increases the load's leverage on the crane and results in the crane's capacity being lowered. This explains why a rough terrain crane can become unstable, even to the point of overturning, when a load is lifted over the front and swung over the side. Be sure to consult the capacity chart before swinging to a less stable area.

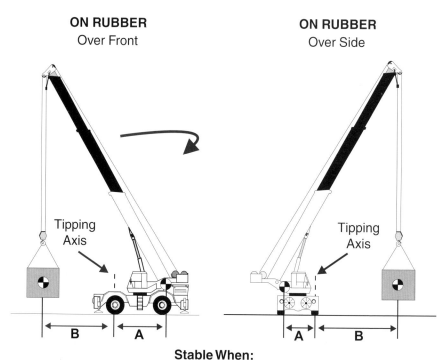

ON RUBBER
Over Front

ON RUBBER
Over Side

Tipping Axis

Tipping Axis

B   A

A   B

Stable When:
(Crane Weight x A) is greater than (Load Weight x B)

A mobile crane is stable when its leverage is greater than the load's leverage.

Leverage and Stability

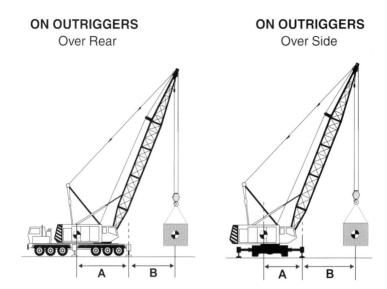

ON OUTRIGGERS
Over Rear

ON OUTRIGGERS
Over Side

Stable When:
(Crane Weight x A) is greater than (Load Weight x B)

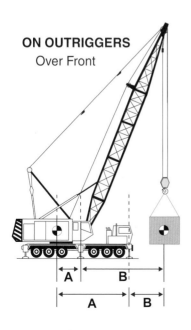

ON OUTRIGGERS
Over Front

Most carrier mounted cranes have their greatest capacity over the rear. When the upper structure is rotated over the side, the capacity will be lower because the distance from the crane's center of gravity to the tipping axis is shortened. Most crane manufacturers do not allow loads to be lifted over the front unless the front stabilizer is extended and set.

Rate of Tipping

As the crane tips, the load's leverage increases and the crane's leverage decreases. This can occur rapidly, making it impossible to recover by setting the load down.

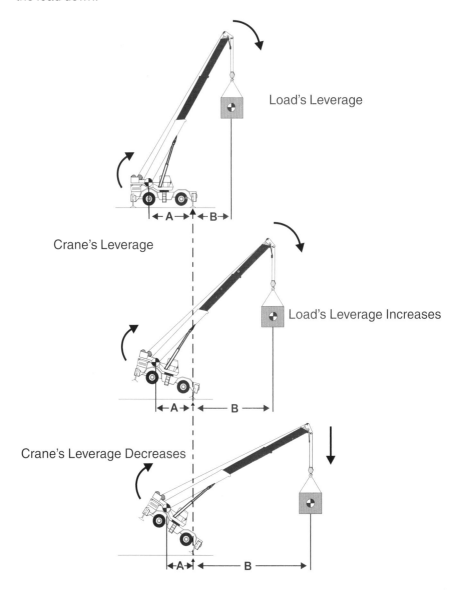

Load's Leverage

Crane's Leverage

Load's Leverage Increases

Crane's Leverage Decreases

Forward Stability Factors

Forward stability is defined as the crane's ability to resist tipping forward. To guard against tipping when developing rated capacities based on stability, manufacturers reduce tipping loads by a percentage established by national standards.

MARGIN OF STABILITY	
Crawler Mounted Cranes	75%
Carrier Mounted Cranes	
on Outriggers Extended	85%
on Rubber or Tires	75%
Commercial Boom Trucks	85%

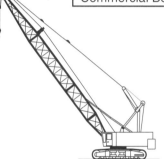

Crawler Cranes
Rated Capacity = Tipping Load x 75%

Carrier Mounted Cranes on Outriggers
Rated Capacity = Tipping Load x 85%

Boom Trucks on Stabilizers
Rated Capacity = Tipping Load x 85%

Carrier Mounted Cranes on Rubber
Rated Capacity = Tipping Load x 75%

Backward Stability Factors

Concern is normally directed toward the crane tipping forward; however, mobile cranes can tip and even overturn in the backward direction. Backward stability is the crane's ability to resist overturning backwards while in an unloaded condition.

Backward stability margins are based on the following general conditions:
- Over side.
- Level within 1% on firm surface.
- All fuel tanks at least 1/2 full.
- Other fluid levels as specified.

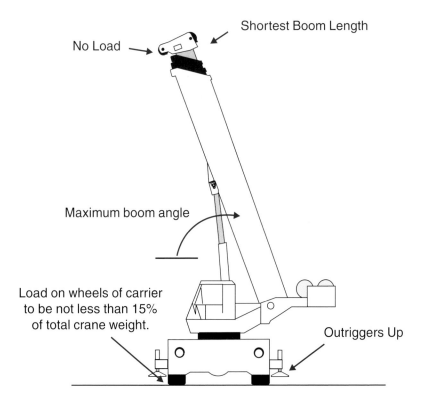

No Load

Shortest Boom Length

Maximum boom angle

Load on wheels of carrier to be not less than 15% of total crane weight.

Outriggers Up

Capacity Limitations

Sufficiently overloading the crane will cause the crane to either tip or fail structurally.

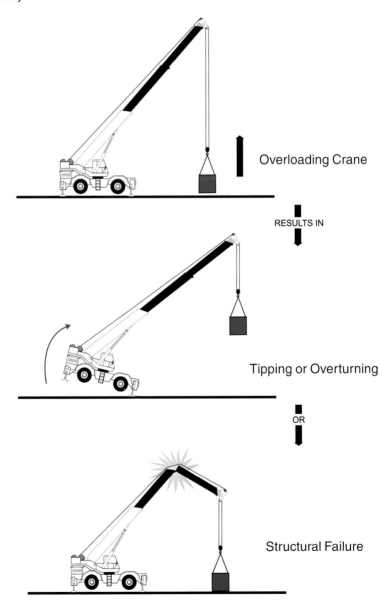

Overloading Crane

RESULTS IN

Tipping or Overturning

OR

Structural Failure

Load Charts

Basis for Rated Capacities

All rated capacities listed in capacity charts are based on either structural strength or stability. These capacities are normally identified by dividing the chart by using a bold line, asterisks or shaded area.

BOLD LINE

RADIUS IN FEET	**Main Boom Length in Feet (Power Pinned Fly Retracted)**							Power Pin Fly & 96 ft.
	40	45	55	65	75	85	96	125
10	120,000 (72)	90,000 (74)	87,300 (77.5)	82,250 (79.5)				
12	98,300 (68.5)	85,400 (71.5)	83,000 (75)	77,400 (78)	60,550 (80)			
15	83,100 (63.5)	79,700 (67)	74,000 (72)	70,500 (75)	55,050 (78)	48,850 (79.5)	33,500 (81.5)	
20	65,550 (54.5)	60,550 (60)	58,700 (66)	55,250 (70.5)	47,250 (74)	41,600 (76)	33,500 (78.5)	21,000 (81.5)
25	46,000 (44.5)	46,000 (51.5)	46,000 (60)	44,100 (65.5)	41,400 (69.5)	36,100 (72.5)	33,000 (75)	21,000 (79)
30	33,300 (31)	33,300 (42)	33,300 (53.5)	33,300 (60.5)	33,300 (65.5)	31,300 (69)	28,150 (72)	19,050 (76.5)
35		24,200 (30)	24,200 (46)	24,200 (55)	24,200 (61)	24,200 (65)	23,800 (68.5)	16,800 (74.5)
40	See Warning Note 16		18,050 (38)	18,050 (49)	18,050 (56.5)	18,050 (61)	18,050 (65.5)	15,000 (72)
45			13,700 (26.5)	13,700 (42.5)	13,700 (51.5)	13,700 (57)	13,700 (62)	13,500 (69.5)
50				10,500 (34.5)	10,500 (46)	10,500 (53)	10,500 (58.5)	12,250 (67)
60					6,100 (32.5)	6,100 (43)	6,100 (50.5)	8,300 (61.5)
70						3,220 (30.5)	3,220 (42)	5,340 (56)
80							1,180 (30.5)	3,250 (50)
90								1,690 (43)

Strength
Stability

ASTERISKS

Boom Radius Feet	Boom Angle Degrees	Boom Point Height Feet	**With Outriggers**	
			Over Front	360' Arc
		83 Ft. BOOM		
20	73.0	88.2	30200 ★	30200 ★
25	69.3	86.4	25900 ★	25900 ★
30	65.6	84.3	27000 ★	27000 ★
35	61.7	81.8	20100 ★	20100 ★
40	57.7	78.8	17800 ★	17800 ★
45	53.5	75.3	16000 ★	16000 ★
50	49.0	71.2	14400	14400
55	44.3	66.4	12000	12000

Strength
Stability

SHADED AREA

Boom Strength	Radius in Feet	Boom Angle Degrees	**Outriggers Set**		Ft. From Boom Point
			Over Side	Over Rear	
	19	80.7	176300	176300	96
	20	80.1	176300	176300	96
	25	7608	160520	160520	95
90 Feet	30	73.5	133780	133780	93
	35	70.2	111770	114720	92
	40	66.8	91440	100320	90
	50	59.6	66560	75740	85
	60	51.9	51870	58880	78

Strength
Stability

Components That Can Fail

Mobile cranes will not only overturn, they will also fail structurally before there is any indication of tipping. Failure is often unexpected when rated capacities are exceeded. This is especially true when tipping is relied upon (seat of the pants operation) while operating in the structural area of the capacity chart. Shown below are some components that can fail.

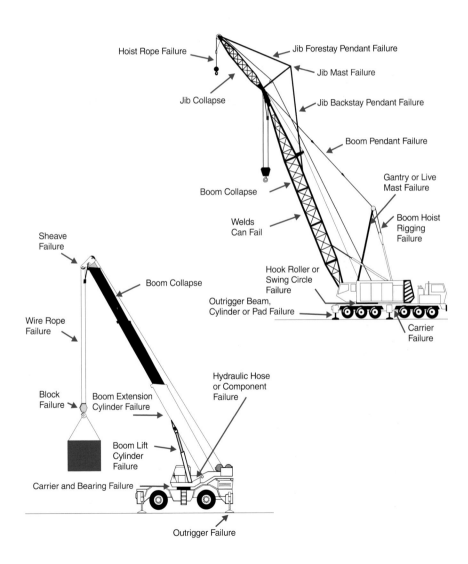

Determining Lifting Capacity

Correctly arriving at the lifting capacity is the main objective of interpreting and applying the load chart. To determine the crane's lifting capacity there are certain steps or procedures that must be followed.

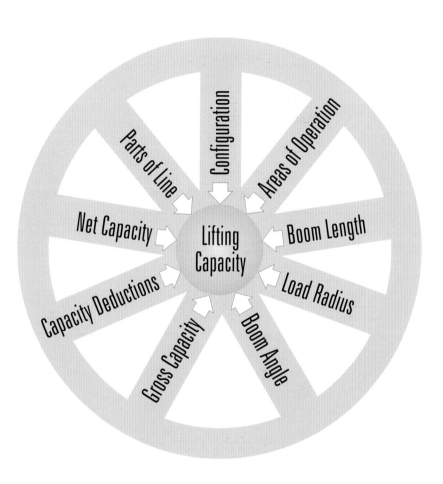

Configuration of Crane Mounting

Selecting the correct capacity chart will be determined by how the outriggers, stabilizers or crawlers are configured (used) on the crane mounting.

The crane must also be assembled per the manufacturer's requirements for the specific load chart used. Some of the most common requirements are: Boom type and sequence, counterweights, gantry position, etc.

Carrier-Mounted

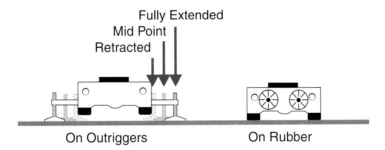

On Outriggers   On Rubber

Crawler-Mounted

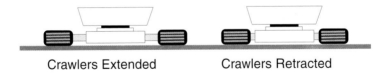

Crawlers Extended   Crawlers Retracted

Truck-Mounted

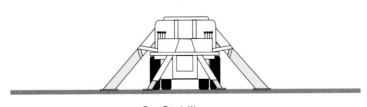

On Stabilizers

Lifting "on rubber" is not normally allowed on boom trucks. Be sure to consult the load chart before operating.

Areas of Operation

Areas of operation, or *quadrants of operation*, refer to a particular part of a crane's working circle. The size of each area may differ from model to model and from manufacturer to manufacturer. Some crane load charts are valid for full 360° operation. Before operating, consult the specific load chart for crane.

Typical quadrants are:

- Over the side
- Over the rear
- Over the front

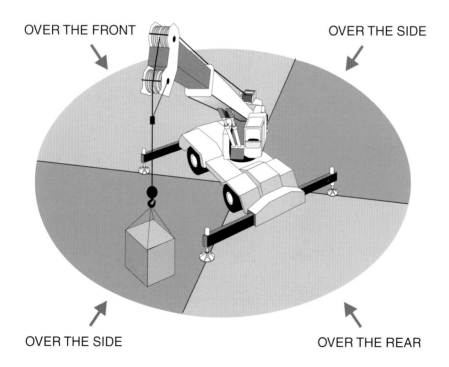

OVER THE FRONT

OVER THE SIDE

OVER THE SIDE

OVER THE REAR

Areas of Operation
Capacity Differences

The capacity of a crane can change depending on the part of the working circle in which it is operating. Many carrier-mounted cranes, for example, can lift more over the rear than they can over the side or front. This is especially true if outriggers are not being used.

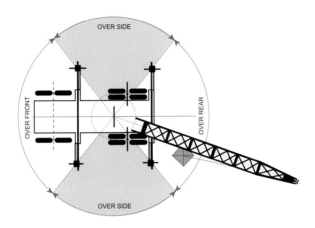

This change in capacity between different *areas of operation* is particularly important when a lift requires a crane to move a load from one area of its working circle to another.

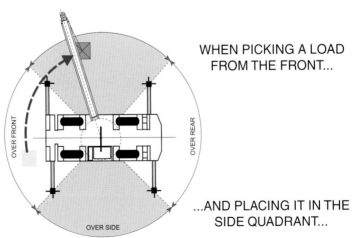

WHEN PICKING A LOAD
FROM THE FRONT...

...AND PLACING IT IN THE
SIDE QUADRANT...

...ALWAYS USE THE LOAD CHART THAT GOVERNS THE WEAKER
QUADRANT (in this case, the side).

Areas of Operation
Carrier-Mounted Cranes

Some mobile cranes, especially older ones, have the quadrants divided differently than those shown. BE SURE to consult the load chart before operating.

WITH OUTRIGGERS FULLY EXTENDED

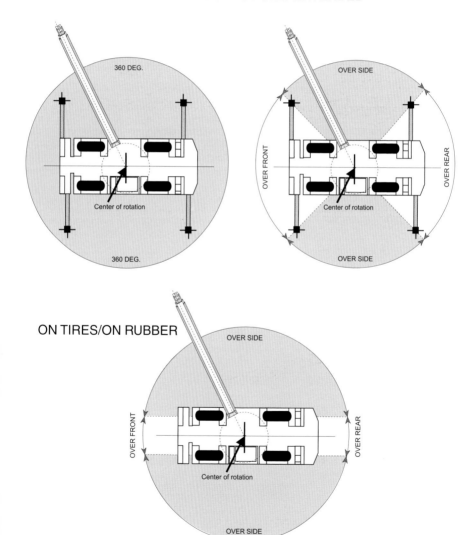

32

Areas of Operation
Carrier-Mounted Cranes

For improved front stability some carrier-mounted cranes are equipped with a front stabilizer (or jack). Do not lift in the front quadrant unless the crane manufacturer approves.

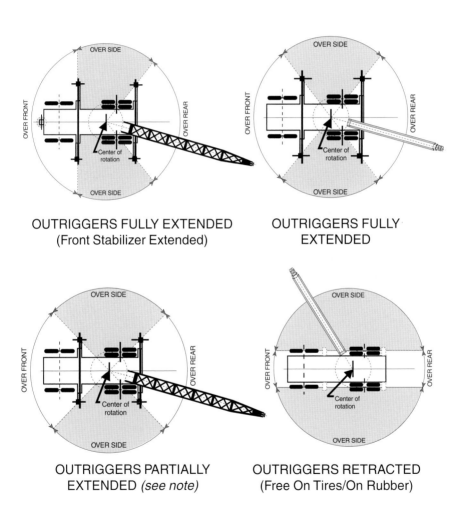

OUTRIGGERS FULLY EXTENDED
(Front Stabilizer Extended)

OUTRIGGERS FULLY EXTENDED

OUTRIGGERS PARTIALLY EXTENDED *(see note)*

OUTRIGGERS RETRACTED
(Free On Tires/On Rubber)

NOTE: Outriggers, when used, must always be FULLY extended unless the manufacturer specifically indicates the crane has been designed for operation with outriggers at intermediate positions, and has provided a load chart for this configuration.

Areas of Operation
Crawler Cranes

There are two types of *areas of operation* for center-mounted crawler cranes: those based on the center of rotation of the upperworks, and those based on the center line of the tracks.

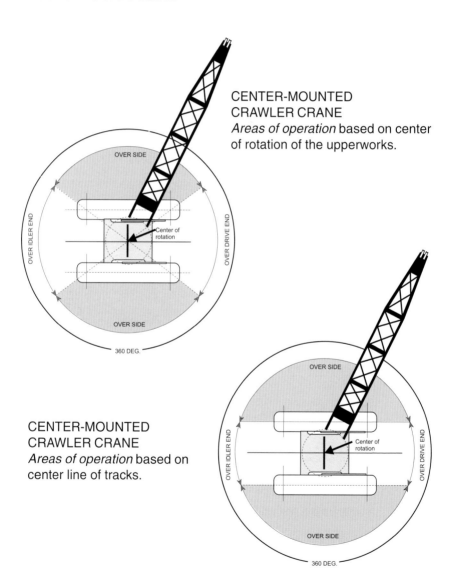

CENTER-MOUNTED
CRAWLER CRANE
Areas of operation based on center of rotation of the upperworks.

CENTER-MOUNTED
CRAWLER CRANE
Areas of operation based on center line of tracks.

Areas of Operation
Crawler Cranes

Crawler cranes may also be offset-mounted rather than center-mounted, in which case their *areas of operation* change accordingly.

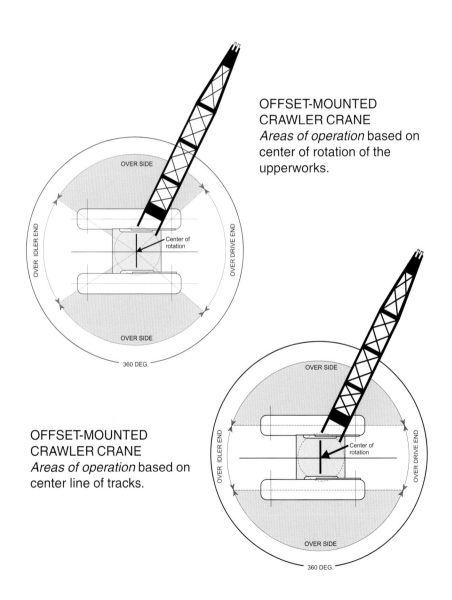

OFFSET-MOUNTED CRAWLER CRANE
Areas of operation based on center of rotation of the upperworks.

OFFSET-MOUNTED CRAWLER CRANE
Areas of operation based on center line of tracks.

Areas of Operation
Commercial Truck-Mounted Cranes

Capacities on boom trucks are normally much lower than other types of mobile cranes and can drop off considerably when swinging loads from one quadrant to another. Be sure to interpret the capacity charts correctly.

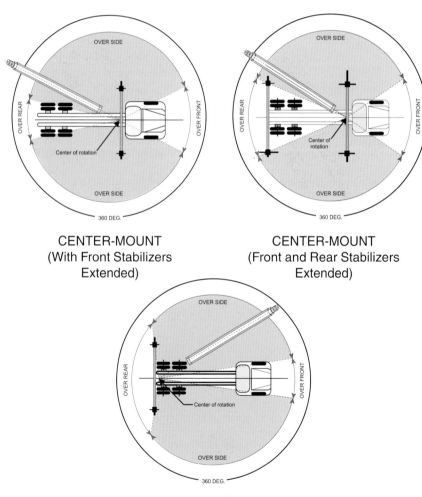

CENTER-MOUNT
(With Front Stabilizers
Extended)

CENTER-MOUNT
(Front and Rear Stabilizers
Extended)

REAR-MOUNT
(Rear Stabilizers Extended)

Boom Length

The meaning of the term *"boom length"* in load charts is not always the same. Sometimes the term includes the boom extension, other times it does not. Be sure to understand what the load chart means by *"boom length"*. Normally it is as shown below:

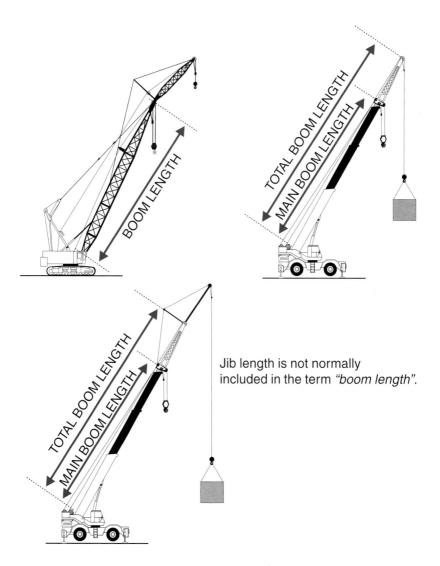

Jib length is not normally included in the term *"boom length"*.

Exact Boom Length in Chart

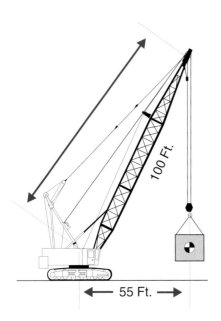

Boom Lgth.: Feet	Oper. Rad.: Feet	Bm. Ang.: Deg.	Boom Point: Elev.	Capacity: Crawlers Retracted	Capacity: Crawlers Extended
	19	81.4	105.9	286,100B	331,400
	20	80.8	105.7	262,300B	318,900
	22	79.6	105.4	224,700B	269,200
	24	78.5	105.0	196,300B	232,500
	26	77.3	104.5	174,000B	204,400
	28	76.1	104.1	156,100B	182,200
	30	74.9	103.6	141,400	164,100
	32	73.7	103.0	129,100	149,200
	34	72.5	102.4	118,700	136,700
	36	71.3	101.7	109,700	125,900
100	38	70.1	101.0	101,900	116,700
	40	68.9	100.3	95,100	108,700
	45	65.8	98.2	81,200	92,400
	50	62.6	95.8	70,600	80,100
	55	59.3	93.0	62,200	70,500
	60	55.9	89.8	35,500	62,800
	65	52.4	86.2	49,900	56,400
	70	48.7	82.1	45,200	51,100
	75	44.8	77.4	41,200	46,600
	80	40.5	72.0	37,700	42,700
	85	35.9	65.6	34,700	39,300
	90	30.7	58.0	32,100	36,300
	95	24.5	48.5	29,800	33,700
	100	16.3	35.0	27,700	31,400

Radius in Feet	Main Boom Length in Feet (Power Pinned Fly Retracted)								
	36	44	52	60	68	76	82	88	114
10	130,000 (67)	106,700 (71.5)	101,600 (74.5)	100,000 (77)	96,700 (79)				
12	120,000 (63)	106,700 (68.5)	101,600 (72)	96,500 (75)	87,850 (77)	84,700 (78.5)			
15	103,450 (57.5)	103,450 (64)	95,300 (68.5)	84,900 (72)	79,180 (74.5)	77,550 (76)	70,250 (77.5)	64,500 (79)	
20	80,650 (47)	80,650 (56.5)	80,650 (62.5)	70,550 (66.5)	64,310 (70)	63,800 (72)	59,400 (74)	55,000 (75.5)	38,750 (80)
25	62,200 (34)	62,200 (48)	62,200 (55.5)	60,150 (61)	54,000 (65.5)	49,700 (67.5)	47,450 (70.5)	45,600 (72)	34,000 (77)
30		48,450 (38)	48,450 (48.5)	48,450 (55.5)	46,650 (65.5)	42,750 (63.5)	40,450 (66.5)	39,150 (68.5)	30,300 (74.5)
35		39,500 (24.5)	39,500 (40.5)	39,500 (49.5)	46,650 (60.5)	37,300 (58.5)	35,200 (62.5)	34,050 (65)	27,250 (71.5)
40			34,400 (30.5)	34,400 (42.5)	34,400 (50)	32,900 (54)	31,000 (58.5)	29,550 (61.5)	24,750 (69)
45			29,250 (14.5)	29,250 (34.5)	29,250 (44)	29,250 (49)	27,500 (54)	26,550 (57.5)	22,650 (66)
50				24,350 (24)	24,350 (37.5)	24,350 (43.5)	24,350 (49.5)	23,750 (53.5)	20,800 (60)
60				17,060 (17.5)	17,060 (30.5)	17,060 (39)	17,060 (44)	17,900 (57)	

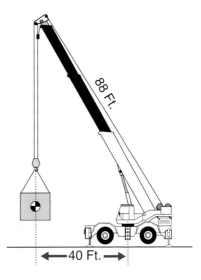

Boom Length Between Chart Listings

Because of the versatility of telescoping boom cranes, it is common when lifting loads for the boom length to be between the boom lengths listed in the load chart.

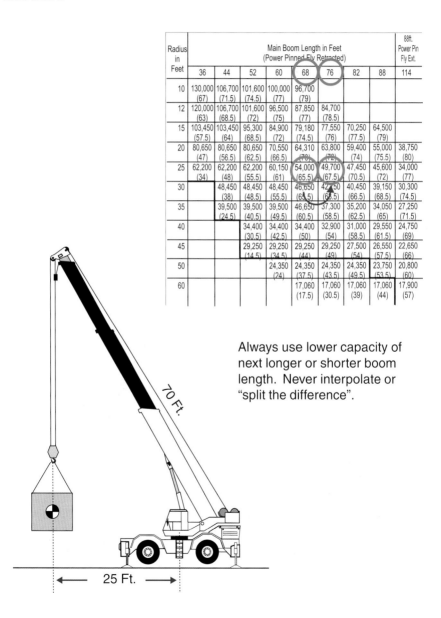

Radius in Feet	Main Boom Length in Feet (Power Pinned Fly Retracted)								88ft. Power Pin Fly Ext.
	36	44	52	60	68	76	82	88	114
10	130,000 (67)	106,700 (71.5)	101,600 (74.5)	100,000 (77)	96,700 (79)				
12	120,000 (63)	106,700 (68.5)	101,600 (72)	96,500 (75)	87,850 (77)	84,700 (78.5)			
15	103,450 (57.5)	103,450 (64)	95,300 (68.5)	84,900 (72)	79,180 (74.5)	77,550 (76)	70,250 (77.5)	64,500 (79)	
20	80,650 (47)	80,650 (56.5)	80,650 (62.5)	70,550 (66.5)	64,310 (70)	63,800 (72)	59,400 (74)	55,000 (75.5)	38,750 (80)
25	62,200 (34)	62,200 (48)	62,200 (55.5)	60,150 (61)	54,000 (65.5)	49,700 (67.5)	47,450 (70.5)	45,600 (72)	34,000 (77)
30		48,450 (38)	48,450 (48.5)	48,450 (55.5)	46,650 (63.5)	42,750 (65.5)	40,450 (66.5)	39,150 (68.5)	30,300 (74.5)
35		39,500 (24.5)	39,500 (40.5)	39,500 (49.5)	46,650 (60.5)	37,300 (58.5)	35,200 (62.5)	34,050 (65)	27,250 (71.5)
40			34,400 (30.5)	34,400 (42.5)	34,400 (50)	32,900 (54)	31,000 (58.5)	29,550 (61.5)	24,750 (69)
45			29,250 (14.5)	29,250 (34.5)	29,250 (44)	29,250 (49)	27,500 (54)	26,550 (57.5)	22,650 (66)
50				24,350 (24)	24,350 (37.5)	24,350 (43.5)	24,350 (49.5)	23,750 (53.5)	20,800 (60)
60					17,060 (17.5)	17,060 (30.5)	17,060 (39)	17,060 (44)	17,900 (57)

Always use lower capacity of next longer or shorter boom length. Never interpolate or "split the difference".

70 Ft.

25 Ft.

Load Radius

The *load radius* is the horizontal distance from the center of rotation to the center line of the hook or center of gravity of the load when lifted.

Boom Length	Radius in Feet	Boom Angle Degrees	Capacity outriggers Extended
	30	73.0	130,400
	32	71.6	118,200
	34	70.3	108,000
	36	68.9	99,300
	38	67.5	91,900
	40	66.2	85,300
	45	62.6	72,300
90 Feet	50	59.0	62,400
	55	55.2	54,700
	60	51.2	48,600
	65	47.0	43,500
	70	42.5	39,300
	75	37.5	35,700
	80	31.9	32,600
	85	25.3	29,900
	90	16.3	27,600

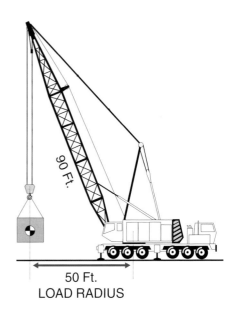

50 Ft.
LOAD RADIUS

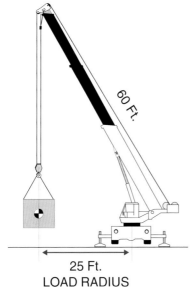

25 Ft.
LOAD RADIUS

Radius in Feet	Main Boom Length in Feet (Power Pinned Fly Retracted)				
	36	44	52	60	68
10	130,000 (67)	106,700 (71.5)	101,600 (74.5)	100,000 (77)	96,700 (79)
12	120,000 (63)	106,700 (68.5)	101,600 (72)	96,500 (75)	87,850 (77)
15	103,450 (57.5)	103,450 (64)	95,300 (68.5)	84,900 (72)	79,180 (74.5)
20	80,650 (47)	80,650 (56.5)	80,650 (62.5)	70,550 (60.5)	64,310 (70)
25	62,200 (34)	62,200 (48)	62,200 (55.5)	60,150 (61)	54,000 (65.5)
30		48,450 (38)	48,450 (48.5)	48,450 (55.5)	46,650 (60.5)
35		39,500 (24.5)	39,500 (40.5)	39,500 (49.5)	39,500 (55.5)
40			31,220 (30.5)	31,220 (42.5)	31,220 (50)
45			24,800 (14.5)	24,800 (34.5)	24,800 (44)

Load Radius

Unless directed by the manufacturer, use the load radius rather than boom angle when determining gross capacity.

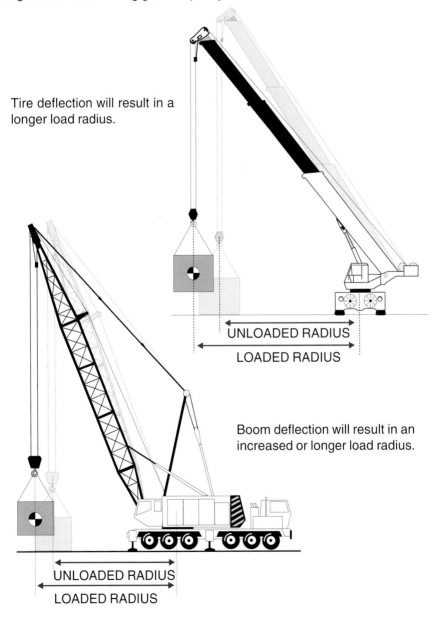

Tire deflection will result in a longer load radius.

UNLOADED RADIUS

LOADED RADIUS

Boom deflection will result in an increased or longer load radius.

UNLOADED RADIUS

LOADED RADIUS

Load Radius Between Chart Listings

Use next longer radius when *load radius* is between the radii listed in the capacity chart. Never interpolate or "split the difference" of the values listed.

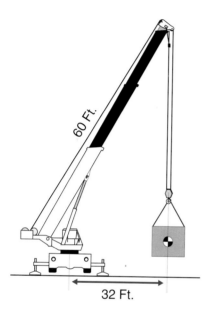

32 Ft.

Radius in Feet	Main Boom Length in Feet (Power Pinned Fly Retracted)				
	36	44	52	60	68
10	130,000 (67)	106,700 (71.5)	101,600 (74.5)	100,000 (77)	96,700 (79)
12	120,000 (63)	106,700 (68.5)	101,600 (72)	96,500 (75)	87,850 (77)
15	103,450 (57.5)	103,450 (64)	95,300 (68.5)	84,900 (72)	79,180 (74.5)
20	80,650 (47)	80,650 (56.5)	80,650 (62.5)	70,550 (66.5)	64,310 (70)
25	62,200 (34)	62,200 (48)	62,200 (55.5)	60,150 (61)	54,000 (65.5)
30		48,450 (38)	48,450 (48.5)	48,450 (55.5)	46,650 (60.5)
35		39,500 (24.5)	39,500 (40.5)	39,500 (49.5)	39,500 (55.5)
40			31,220 (30.5)	31,220 (42.5)	31,220 (50)
45			24,800 (14.5)	24,800 (34.5)	24,800 (44)

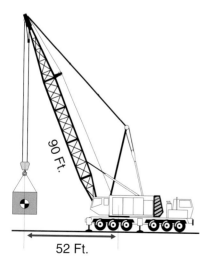

52 Ft.

Boom Length	Radius in Feet	Boom Angle Degrees	Capacity outriggers Extended
	30	73.0	130,400
	32	71.6	118,200
	34	70.3	108,000
	36	68.9	99,300
	38	67.5	91,900
	40	66.2	85,300
	45	62.6	72,300
90 Feet	50	59.0	62,400
	55	55.2	54,700
	60	51.2	48,600
	65	47.0	43,500
	70	42.5	39,300
	75	37.5	35,700
	80	31.9	32,600
	85	25.3	29,900
	90	16.3	27,600

Boom Angle

The *boom angle* on telescoping boom cranes is the angle between the center line of the base boom and the horizontal after the load is lifted.

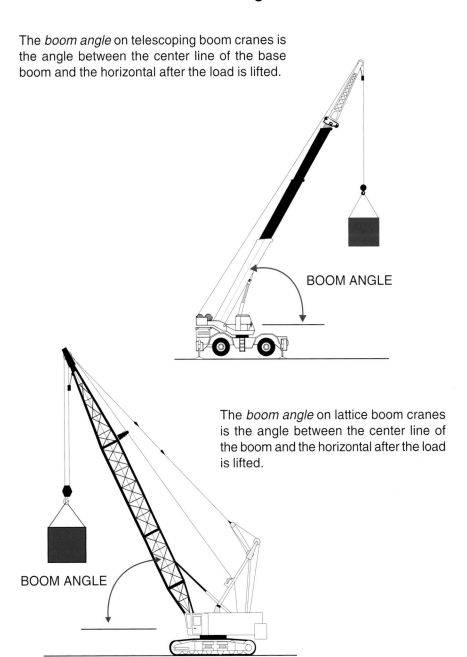

BOOM ANGLE

The *boom angle* on lattice boom cranes is the angle between the center line of the boom and the horizontal after the load is lifted.

BOOM ANGLE

Boom Angle

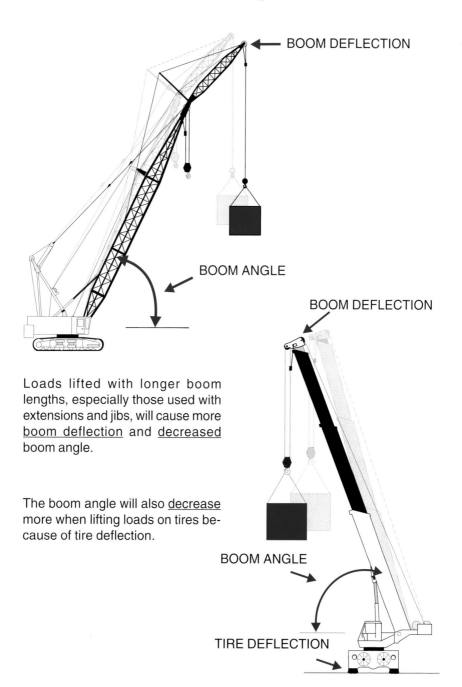

BOOM DEFLECTION

BOOM ANGLE

BOOM DEFLECTION

Loads lifted with longer boom lengths, especially those used with extensions and jibs, will cause more boom deflection and decreased boom angle.

The boom angle will also decrease more when lifting loads on tires because of tire deflection.

BOOM ANGLE

TIRE DEFLECTION

Boom Angle Between Chart Listings

When boom angle is between the angles listed in the capacity chart, always use next lower boom angle. Do not interpolate.

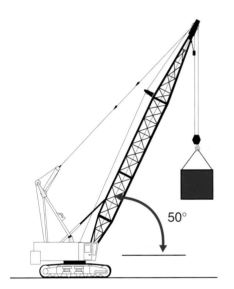

50°

Boom Lgth.: Feet	Oper. Rad.: Feet	Bm. Ang.: Deg.	Boom Point: Elev.	Capacity: Crawlers Retracted	Capacity: Crawlers Extended
19	81.4	105.9	286,100B	331,400
20	80.8	105.7	262,300B	318,900
22	79.6	105.4	224,700B	269,200
24	78.5	105.0	196,300B	232,500
26	77.3	104.5	174,000B	204,400
28	76.1	104.1	156,100B	182,200
30	74.9	103.6	141,400	164,100
32	73.7	103.0	129,100	149,200
34	72.5	102.4	118,700	136,700
36	71.3	101.7	109,700	125,900
38	70.1	101.0	101,900	116,700
40	68.9	100.3	95,100	108,700
45	65.8	98.2	81,200	92,400
50	62.6	95.8	70,600	80,100
55	59.3	93.0	62,200	70,500
60	55.9	89.8	35,500	62,800
65	52.4	86.2	49,900	56,100
70	48.7	82.1	45,200	51,100
75	44.9	77.4	41,200	46,600
80	40.5	72.0	37,700	42,700
85	35.9	65.6	34,700	39,300
90	30.7	58.0	32,100	36,300
95	24.5	48.5	29,800	33,700
100	16.3	35.0	27,700	31,400

Boom Lgth.: 100

Radius in Feet	Main Boom Length in Feet (Power Pinned Fly Retracted)					88ft. Power Pin Fly Ext.	32ft. Ext. & 88ft.	32ft. Ext. & 114ft.
	60	68	76	82	88	114	120	146
10	100,000 (77)	96,700 (79)						
12	96,500 (75)	87,850 (77)	84,700 (78.5)					
15	84,900 (72)	79,180 (74.5)	77,550 (76)	70,250 (77.5)	64,500 (79)			
20	70,550 (66.5)	64,310 (70)	63,800 (72)	59,400 (74)	55,000 (75.5)	38,750 (80)	23,600 (79.5)	
25	60,150 (61)	54,000 (65.5)	49,700 (67.5)	47,450 (70.5)	45,600 (72)	34,000 (77)	21,300 (77)	22,500 (80)
30	48,450 (55.5)	46,650 (65.5)	42,750 (63.5)	40,450 (66.5)	39,150 (68.5)	30,300 (74.5)	19,500 (74.5)	20,400 (78.5)
35	39,500 (49.5)	46,650 (60.5)	37,300 (58.5)	35,200 (62.5)	34,050 (65)	27,250 (71.5)	17,950 (72)	18,000 (76.5)
40	34,400 (42.5)	34,400 (50)	32,900 (54)	31,000 (58.5)	29,550 (61.5)	24,750 (69)	16,600 (74.5)	16,000 (74.5)
45	29,250 (34.5)	29,250 (44)	29,250 (49)	27,500 (54)	26,550 (57.5)	22,650 (66)	15,500 (66.5)	14,620 (72.5)
50	24,350 (24)	24,350 (37.5)	24,350 (43.5)	24,350 (49.5)	23,750 (53.5)	20,800 (60)	15,500 (64)	13,730 (70)
60		17,060 (17.5)	17,060 (30.5)	17,060 (39)	17,060 (44)	17,900 (57)	12,850 (58.5)	11,450 (66)

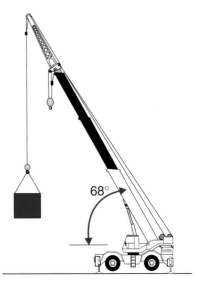

68°

45

Gross Capacity

Gross capacities, which are sometimes called rated capacities, are listed in the capacity charts for the appropriate boom length, boom angle and radius.

Gross or rated capacities <u>are not</u> the maximum loads or objects that can be lifted.

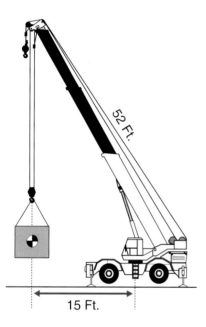

15 Ft.

Radius in Feet	Main Boom Length in Feet (Power Pinned Fly Retracted)				
	36	44	52	60	68
10	130,000 (67)	106,700 (71.5)	101,600 (74.5)	100,000 (77)	96,700 (79)
12	120,000 (63)	106,700 (68.5)	101,600 (72)	96,500 (75)	87,850 (77)
15	103,450 (57.5)	103,450 (64)	95,300 (68.5)	84,900 (72)	79,180 (74.5)
20	80,650 (47)	80,650 (56.5)	80,650 (62.5)	70,550 (66.5)	64,310 (70)
25	62,200 (34)	62,200 (48)	62,200 (55.5)	60,150 (61)	54,000 (65.5)
30		48,450 (38)	48,450 (48.5)	48,450 (55.5)	46,650 (65.5)
35		39,500 (24.5)	39,500 (40.5)	39,500 (49.5)	46,650 (60.5)

Boom Angle	5° OFFSET		17° OFFSET		30° OFFSET	
	Radius (ref.) ft.	Caps. (lbs.)	Radius (ref.) ft.	Caps. (lbs.)	Radius (ref.) ft.	Caps. (lbs.)
80°	32.6	10,000	38.1	8,450	42.3	6,430
75	47.1	8,720	52.2	7,430	56.4	5,870
70	61.2	7,430	65.8	6,520	70.1	5,510
65	74.8	6,330	78.9	5,600	82.8	4,770
60	87.8	5,230	91.4	4,680	94.9	4,130
55	100.2	3,270	103.2	3,040	106.2	3,040
50	111.8	2,210	114.2	2,100	116.8	2,100
45	122.3	1,340	124.4	1,220	126.4	1,220

Gross Capacity

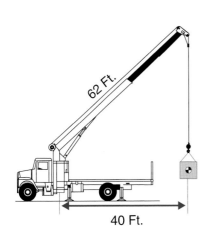

62 Ft.

40 Ft.

BOOM LOAD RATINGS

LOAD RADIUS (FEET)	LOADED BOOM ANGLE	44 FT BOOM (LBS)	LOADED BOOM ANGLE	53 FT BOOM (LBS)	LOADED BOOM ANGLE	62 FT BOOM (LBS)
4.5						
8						
10	77.5	12,100				
12	75.0	10,300	78.5	9,750		
14	72.0	9,400	76.0	8,500	78.5	8,100
16	70.0	8,400	73.5	7,350	76.0	7,000
20	63.5	6,650	69.0	6,050	72.5	5,850
25	56.0	5,520	63.5	5,000	67.0	4,450
30	47.5	4,350	56.5	4,200	62.0	3,900
35	37.0	3,700	49.5	3,650	56.5	3,400
40	23.0	3,100	42.0	3,200	51.0	3,000
45			32.0	2,700	44.0	2,600
50			18.5	2,200	37.0	2,250
55					27.0	1,900

JIB LOAD RATINGS

LOAD RADIUS (FEET)	LOADED BOOM ANGLE	24 FT. JIB (LBS)
16	80	3,000
20	77	2,700
25	74	2,450
30	71	2,100
35	67	1,750
40	63	1,450
45	59	1,200
50	55	950
55	51	800
60	46	600
65	41	550

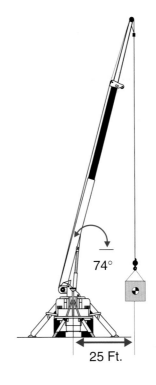

74°

25 Ft.

Capacity Deductions

Load handling devices can add a great deal of weight to the crane and lower crane capacities considerably. All load handling devices, which includes all rigging, must be considered as part of the load and deductions made from the gross or rated capacity.

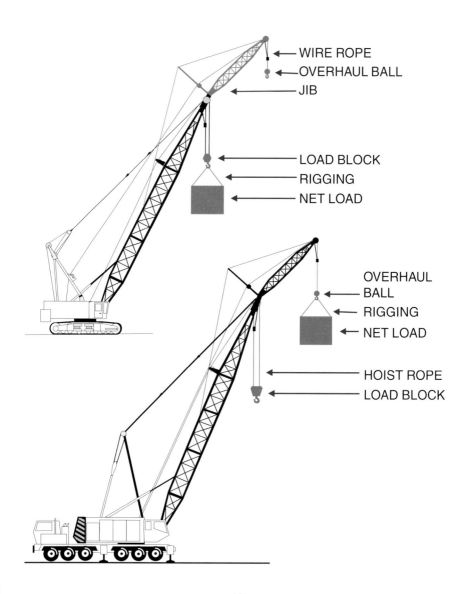

Capacity Deductions

Depending on its location, the actual weight of the load handling device may be lower or even higher than its actual weight. Therefore, when required, it is crucial that devices used for lifting be taken into consideration.

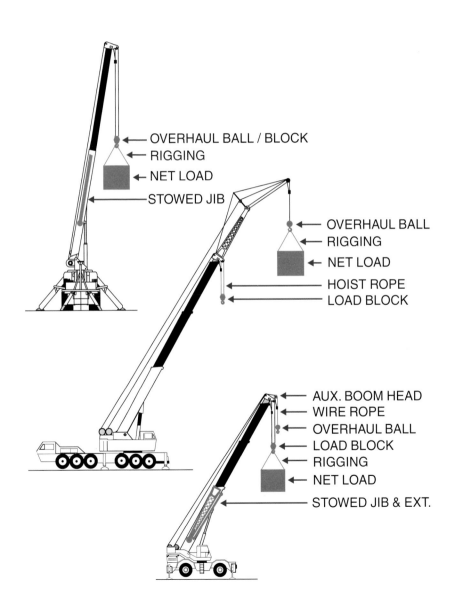

Capacity Deductions

When to Deduct Weight of Crane's Hoist Rope

When manufacturer specifies in the load chart that hoist rope be deducted.

When crane is reeved with more than the minimum parts of line required to lift the load.

When crane is reeved with hoist line not being used to make the lift.

When there are parts of line below ground level.

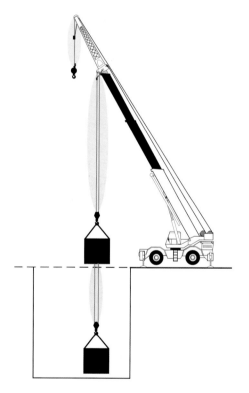

Rope Diameter Inches	Approx. Weight Pounds Per Foot (IWRC)
1/4	.12
5/16	.18
3/8	.26
7/16	.35
1/2	.47
9/16	.60
5/8	.73
3/4	1.06
7/8	1.44
1	1.88
1-1/8	2.34
1-1/4	2.89
1-3/8	3.50
1-1/2	4.16
1-5/8	4.88
1-3/4	5.67
1-7/8	6.50
2	7.39
2-1/8	8.35
2-1/4	9.36
2-3/8	10.40
2-1/2	11.60
2-5/8	12.80
2-3/4	14.00

Unless it is known for certain that the hoist rope weight has been taken into consideration by the crane manufacturer, use the guidelines above.

Net Capacity

Net capacity is Gross capacity minus capacity deductions for load handling devices. Net capacity is the maximum net load or object that can be lifted. The crane's lifting capacity may also be limited by parts of line, hoist rope capacity, overhaul ball or hook block capacity, etc.

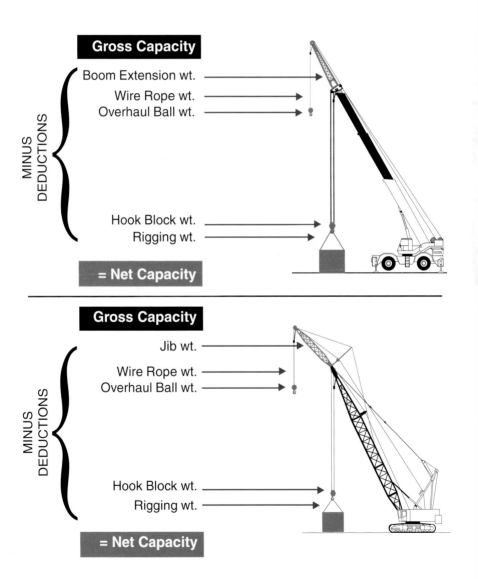

Net Capacity

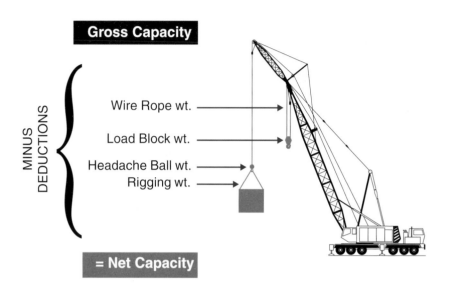

Gross Capacity

MINUS DEDUCTIONS {

Wire Rope wt. ⟶

Load Block wt. ⟶

Headache Ball wt. ⟶
Rigging wt. ⟶

= Net Capacity

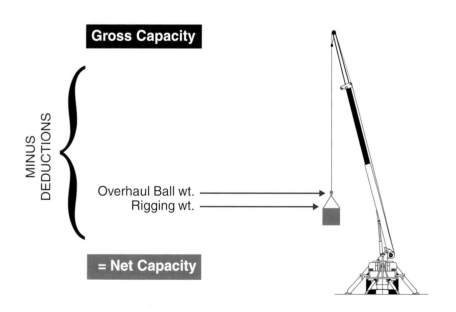

Gross Capacity

MINUS DEDUCTIONS {

Overhaul Ball wt. ⟶
Rigging wt. ⟶

= Net Capacity

Gross Load vs Net Load

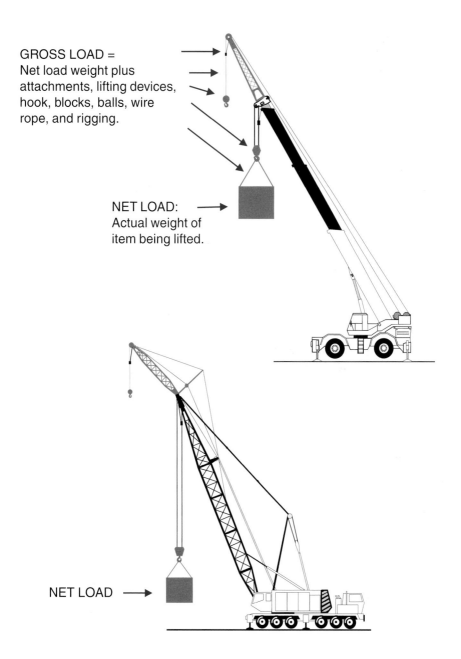

GROSS LOAD =
Net load weight plus
attachments, lifting devices,
hook, blocks, balls, wire
rope, and rigging.

NET LOAD:
Actual weight of
item being lifted.

NET LOAD

Dynamic Loading

Crane capacities are based on static or stationary loads. Forces created by movement of the load and attachments affect crane capacities much the same way as adding additional load weight. Erratic or sudden movement of loads must be avoided. Always operate the crane in a smooth and controlled manner.

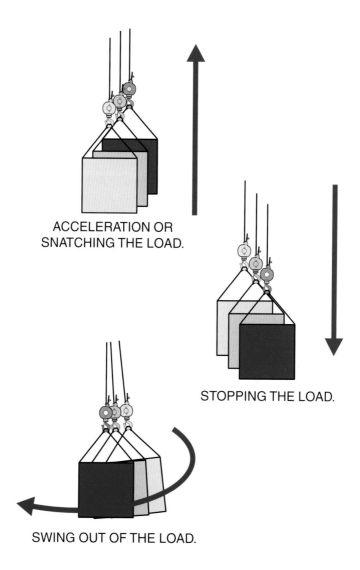

ACCELERATION OR
SNATCHING THE LOAD.

STOPPING THE LOAD.

SWING OUT OF THE LOAD.

Total Load

Total load is the combined forces produced by the actual load weight, lifting attachments and movement of the load and attachments.

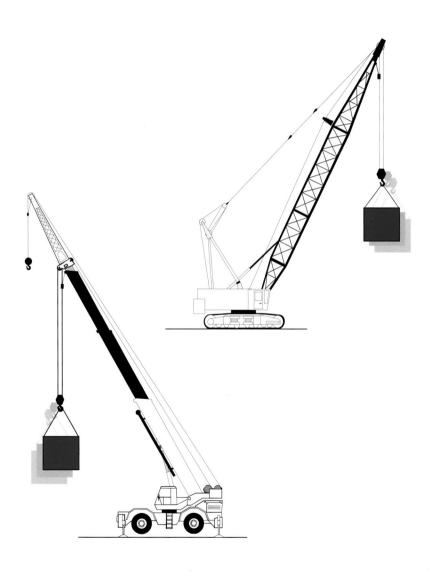

Lifting Capacity

If not properly equipped, the crane's *Lifting Capacity* could be lower than the net capacity and the crane may not be capable of lifting the net load safely. In such cases, capacity is limited by the capacity of the lowest rated component, i.e. hoist rope, hook block, etc. The total load must not exceed the lowest rated capacity of these components. The following are examples:

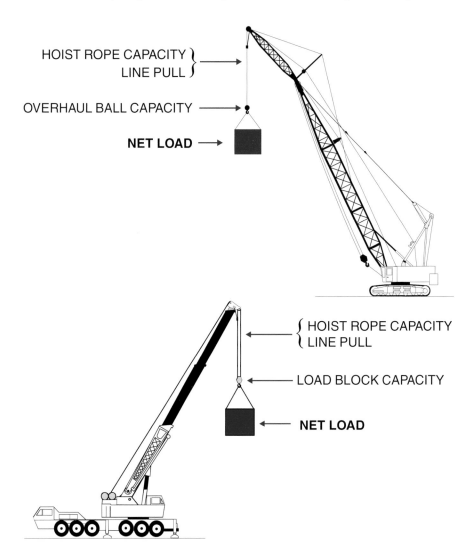

NOTE: Lifting capacity can also be limited by wind.

Parts of Line

PARTS OF LINE = The number of hoist ropes directly supporting the lower load block or overhaul ball.

To avoid breaking the hoist rope and to ensure that the hoist winch can safely lift the load, the load block must be reeved with at least minimum parts of hoist line.

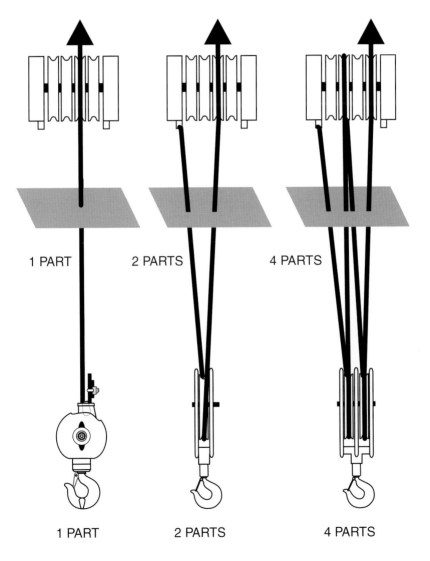

1 PART   2 PARTS   4 PARTS

1 PART   2 PARTS   4 PARTS

Parts of Line

Use the following formula to determine minimum parts of line required to lift the net load: The weight of all devices used to lift the load combined with the actual load weight equals the suspended weight.

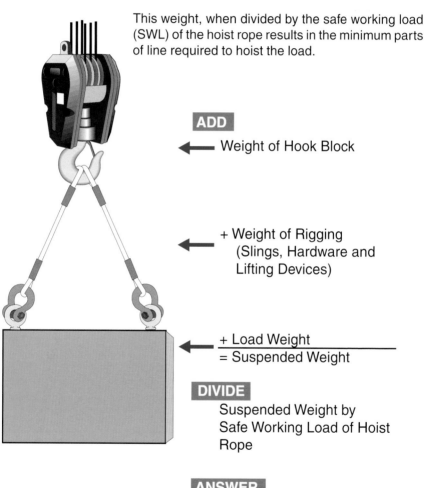

This weight, when divided by the safe working load (SWL) of the hoist rope results in the minimum parts of line required to hoist the load.

ADD

← Weight of Hook Block

← + Weight of Rigging (Slings, Hardware and Lifting Devices)

← + Load Weight

= Suspended Weight

DIVIDE

Suspended Weight by Safe Working Load of Hoist Rope

ANSWER

Minimum Parts of Line Required

Range Diagram

The range diagram provided in the load chart gives a side view of the boom, extension and/or jib downward arcs plus hook elevation and radii.

The diagram can be used to derive approximate distances when planning lifts and determining crane configurations, etc.

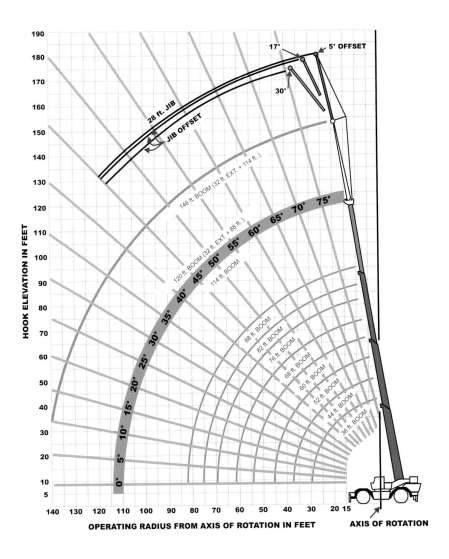

Contents

Crane Not to Specification

Capacity chart ratings apply only to cranes rigged and assembled as stipulated by the crane manufacturer.

OSHA requires that no modification or addition which affects the capacity or safe operation of the crane be made without the manufacturer's written approval.

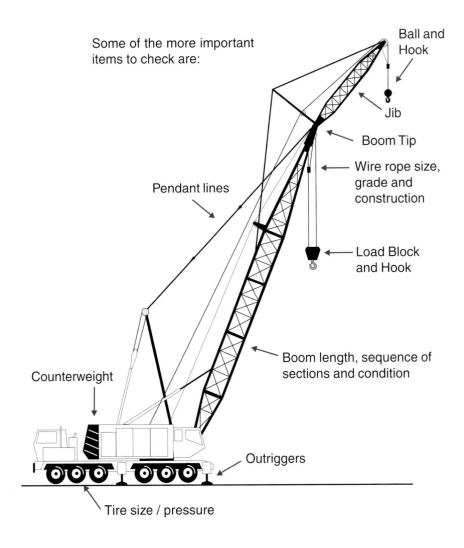

Some of the more important items to check are:

Ball and Hook

Jib

Boom Tip

Wire rope size, grade and construction

Pendant lines

Load Block and Hook

Boom length, sequence of sections and condition

Counterweight

Outriggers

Tire size / pressure

Crane Condition

The boom is one of the most critical components of the crane and must be in good condition at all times.

With telescoping booms check for:

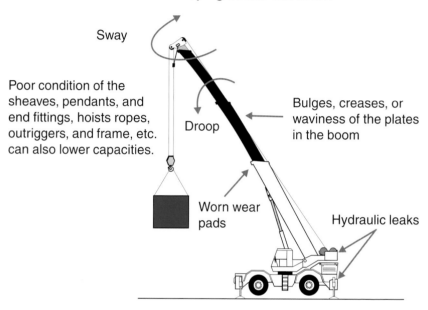

Sway

Poor condition of the sheaves, pendants, and end fittings, hoists ropes, outriggers, and frame, etc. can also lower capacities.

Droop

Bulges, creases, or waviness of the plates in the boom

Worn wear pads

Hydraulic leaks

With lattice booms check:

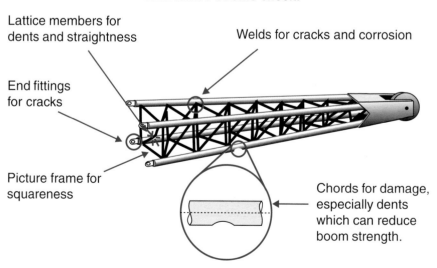

Lattice members for dents and straightness

Welds for cracks and corrosion

End fittings for cracks

Picture frame for squareness

Chords for damage, especially dents which can reduce boom strength.

Crane Not Level

Load chart capacities are generally based on the crane being level within 1% in all directions throughout the lift.

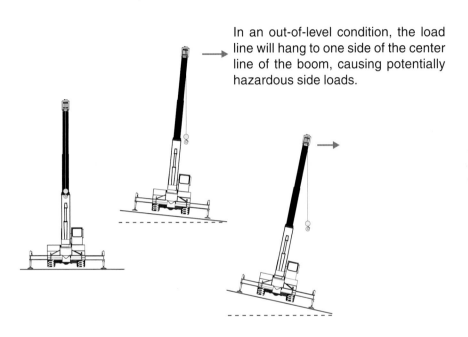

In an out-of-level condition, the load line will hang to one side of the center line of the boom, causing potentially hazardous side loads.

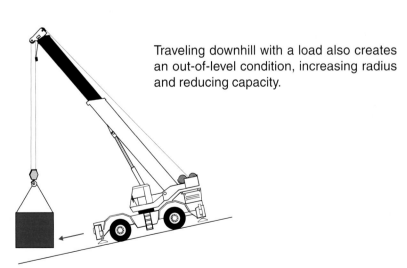

Traveling downhill with a load also creates an out-of-level condition, increasing radius and reducing capacity.

High Winds

The force created by wind can have a devastating effect on the crane. Most crane manufacturers require in the load chart or operator's manual that the rated capacities be reduced when operating in windy conditions. In most cases crane operations should be stopped when the wind exceeds 30 mph.

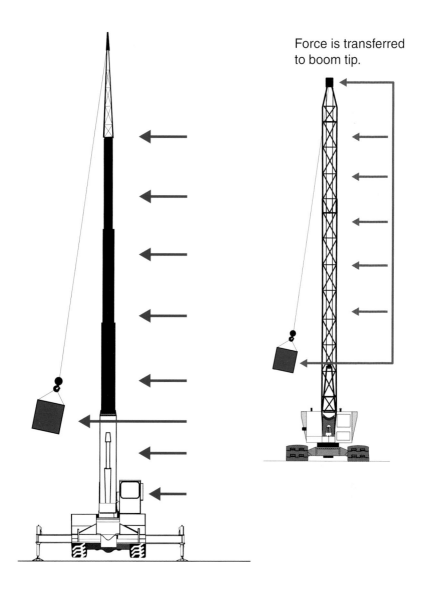

Force is transferred to boom tip.

High Winds

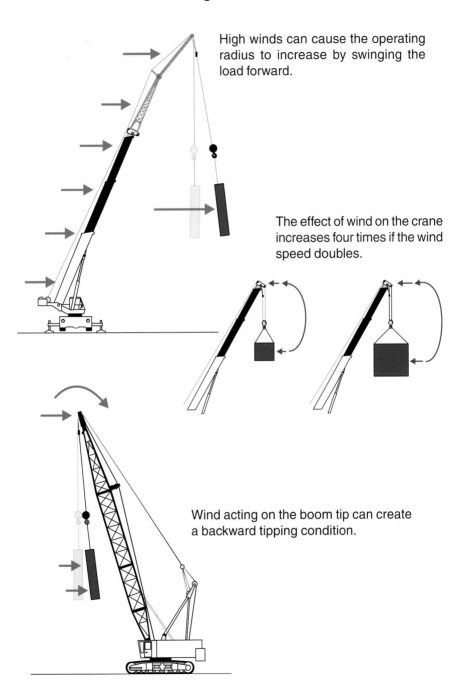

High winds can cause the operating radius to increase by swinging the load forward.

The effect of wind on the crane increases four times if the wind speed doubles.

Wind acting on the boom tip can create a backward tipping condition.

Side Loading

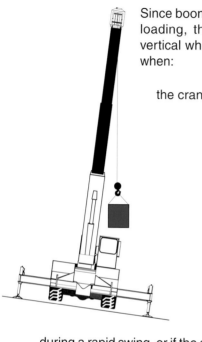

Since booms are only designed to take minimal side loading, the load line should remain reasonably vertical when moving loads. Side loading can occur when:

the crane is out of level, or . . .

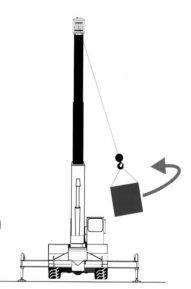

during a rapid swing, or if the swing brake is applied suddenly.

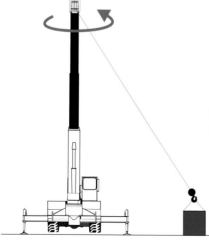

Dragging or pulling loads sideways is not permitted.

Side Loading

Tilt-up operations can also cause side loading if not performed properly.

Side loading can severely affect both lattice and telescoping booms, causing them to buckle.

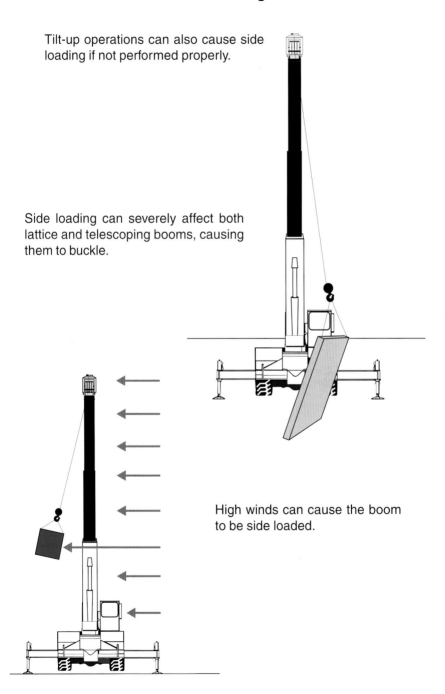

High winds can cause the boom to be side loaded.

Increase in Load Radius

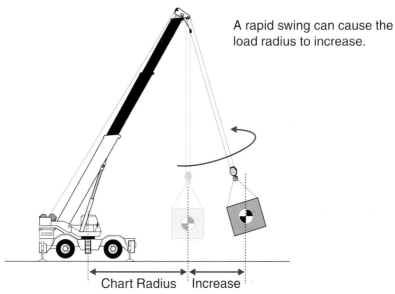

A rapid swing can cause the load radius to increase.

Chart Radius | Increase

As crane starts swing

During swing

After crane stops swing

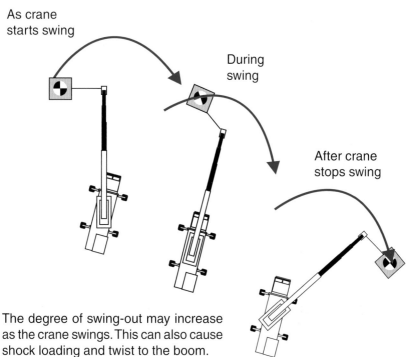

The degree of swing-out may increase as the crane swings. This can also cause shock loading and twist to the boom.

Increase in Load Radius

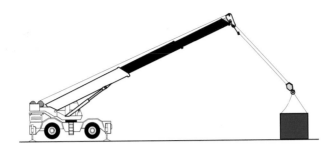

Reaching beyond the vertical extends the load radius and can pull the boom forward.

Swinging a load from over the rear to over the side, or from over the front to over the side, can increase carrier deflection and extend load radius. This is especially noticeable when working on rubber.

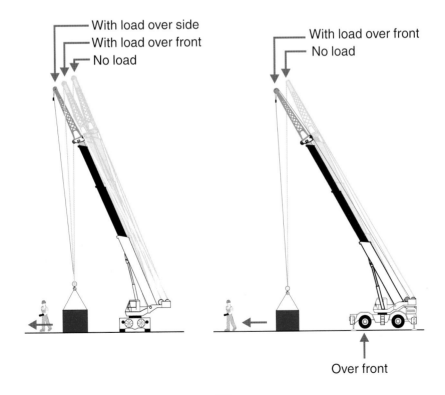

With load over side
With load over front
No load

With load over front
No load

Over front

Increase in Load Radius

Lifting from *inside* the boom tip radius can also cause the load to swing out, increasing the working radius and decreasing capacity.

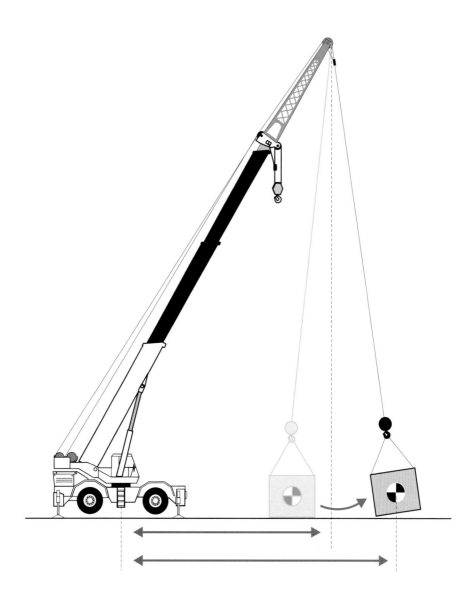

Dynamic Loading

Line Speed Ft. / Min.	Stopping Distance (Ft.)		
	10	6	2
100	0.4%	0.7%	2.2%
200	1.7%	2.9%	8.6%
300	3.9%	6.5%	19.4%
400	6.9%	11.5%	34.5%

Sudden stopping of the load produces hook loads higher than the actual load weight.

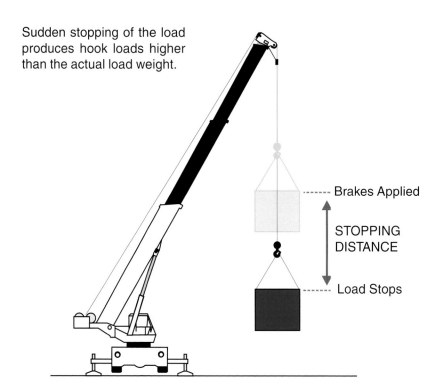

Brakes Applied

STOPPING DISTANCE

Load Stops

Shock Loading

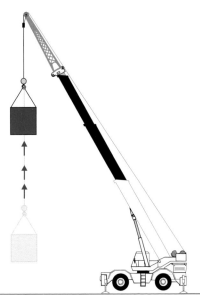

Rapid hoist acceleration produces hook loads higher than the actual load weight.

Sudden release of a load can cause the crane to tip backwards . . .

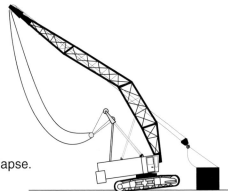

. . . or even collapse.

Shock Loading

Pick and carry operations can subject the carrier and the boom to shock loads.

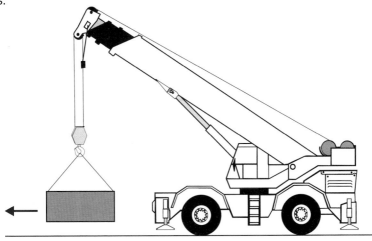

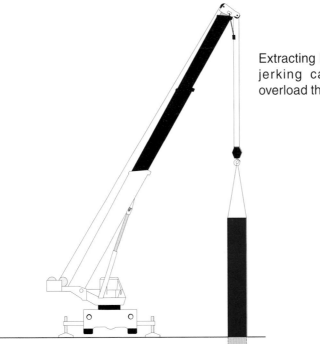

Extracting loads by jerking can also overload the crane.

Shock Loading

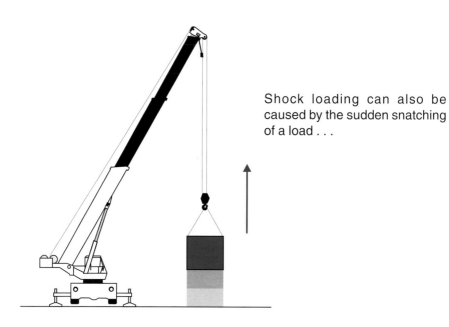

Shock loading can also be caused by the sudden snatching of a load . . .

. . . or the sudden release of a frozen, caught or stuck load.

Either can cause overloading or structural failure.

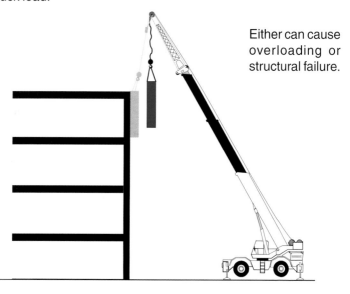

Eccentric Reeving

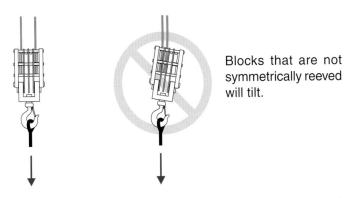

Blocks that are not symmetrically reeved will tilt.

When a hoist line runs on the center sheave, or a sheave beside the center line of the boom, torque is minimized or even eliminated.

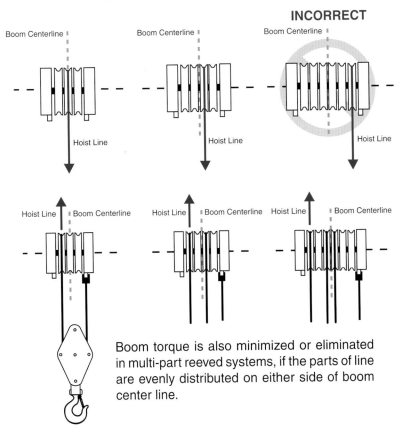

Boom torque is also minimized or eliminated in multi-part reeved systems, if the parts of line are evenly distributed on either side of boom center line.

Duty Cycle Operations

When cranes are used in high speed production operations (duty cycle operations), they are subject to side loading, swing out, and shock loading.

For duty cycle applications, manufacturers usually down-rate their cranes by 20% or provide special load charts.

Examples of typical duty cycle operations are:

High speed, high volume concrete placement

High speed, high volume steel erection

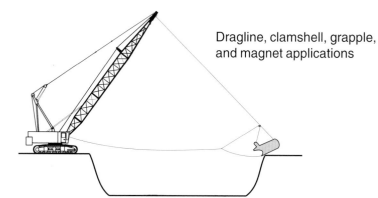

Dragline, clamshell, grapple, and magnet applications

Contents

Frequency of Inspection

OSHA requires all mobile cranes to be inspected regularly.

A <u>frequent inspection</u> (often referred to as the pre-operational inspection) must be made by a competent person before each use of the crane. This inspection is usually the responsibility of the crane operator.

Since the crane must be in a safe operating condition at all times, this inspection essentially continues the whole time the crane is in use.

OSHA also requires a monthly inspection of critical items and a thorough inspection of the entire crane at least annually. Both inspections require that records be kept.

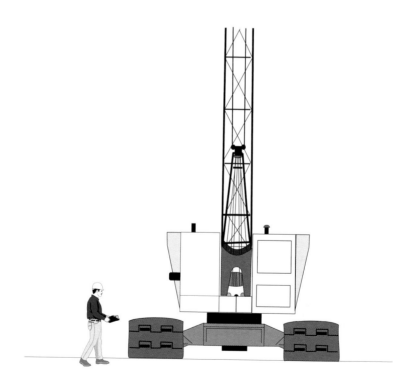

Key Items of Inspection

Using the criteria in OSHA 29 CFR 1926.550 and ANSI/ASME B30.5, as well as the guidelines provided by the manufacturer in the crane's operator's manual, the operator must at a minimum inspect the specific items of the crane illustrated below.

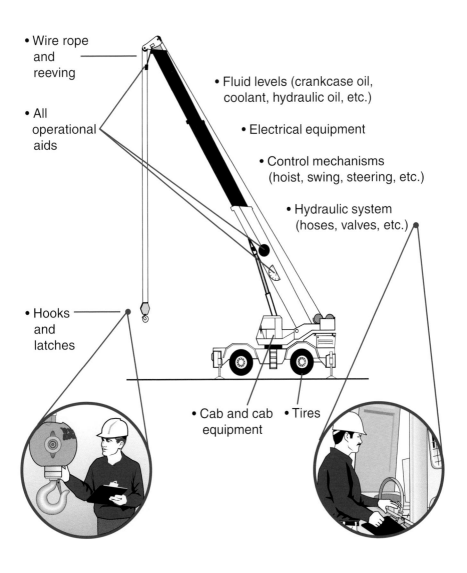

- Wire rope and reeving
- All operational aids
- Fluid levels (crankcase oil, coolant, hydraulic oil, etc.)
- Electrical equipment
- Control mechanisms (hoist, swing, steering, etc.)
- Hydraulic system (hoses, valves, etc.)
- Hooks and latches
- Cab and cab equipment
- Tires

Physical Walk-Round Inspection

The operator should also perform a walk-round inspection of the crane checking for any apparent deficiencies. Other areas to inspect include the carrier or car body, outriggers or crawlers, gear train, and the upper works.

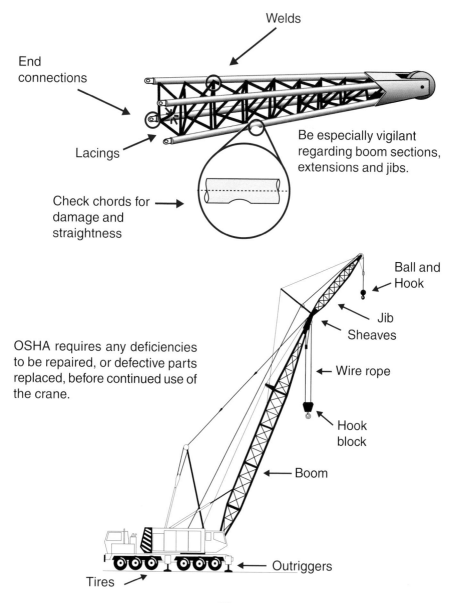

Welds

End connections

Lacings

Check chords for damage and straightness

Be especially vigilant regarding boom sections, extensions and jibs.

Ball and Hook

Jib Sheaves

OSHA requires any deficiencies to be repaired, or defective parts replaced, before continued use of the crane.

Wire rope

Hook block

Boom

Outriggers

Tires

Wire Rope Inspection

All running ropes (main, auxiliary and boom hoists) should be visually inspected at least daily. The inspection should cover at least the rope expected to be used during the day's operations.

During the inspection look for:

BIRDCAGING

- evidence of heat damage

- distortion (kinking, crushing, birdcaging, strand displacement, and core protrusion)

- corrosion

BROKEN WIRES

- broken or cut strands

- broken wires

- core failure (in rotation resistant ropes)

CORE PROTUSION

Wire Rope Inspection

Also look for:

- reduction of rope diameter possibly caused by core failure, corrosion, or wear

- corroded or broken wires at end connections

- damaged, worn, or improperly installed end connections

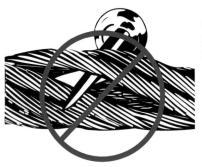

Only the surface wires need be inspected. <u>Under no circumstances should the rope be opened up.</u>

Wear and damage to wire rope is more likely at contact points such as cross-over points, sheaves for running ropes and especially equalizer sheaves.

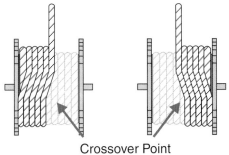

Crossover Point

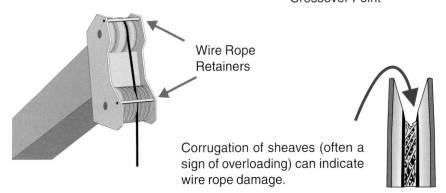

Wire Rope Retainers

Corrugation of sheaves (often a sign of overloading) can indicate wire rope damage.

When to Replace Wire Rope

OSHA also requires running ropes (main and auxiliary hoists, boom hoist) to be taken out of service when any of the following conditions exist:

- 6 randomly distributed broken wires in one lay length, or 3 broken wires in one strand in one lay length.

  The term lay length refers to the distance it takes for a strand to make a complete revolution around the core.

Lay Length

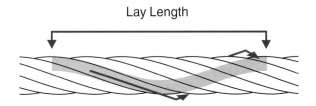

- Wear of one-third the original diameter of outside individual wires.

- Evidence of heat damage.

- Reductions from nominal diameter of the rope *(see table).*

NOMINAL DIAMETER OF ROPE	MAX. ALLOWABLE REDUCTION
up to 5/16 in.	1/64 in.
3/8 up to 1/2 in.	1/32 in.
9/16 up to 3/4 in.	3/64 in.
7/8 up to 1 1/8 in.	1/16 in.
1 1/4 up to 1 1/2 in.	3/32 in.

Other Wire Rope Criteria

Other damage which would require the rope to be removed from service includes:

- In rotation resistant ropes:

  2 randomly distributed broken wires in 6 rope diameters, or

  4 randomly distributed broken wires in 30 rope diameters.

- One outer wire broken at the point of contact with the core of the rope which protrudes or loops out from the rope structure.

- In standing ropes, more than 2 broken wires in one lay in sections beyond end connections, or more than 1 broken wire at an end connection.

This wire rope and end connection must be replaced.

Functional Test

Before each use of the crane, the operator should also run the crane through all its functions for smooth and correct operation. This would include boom up/down; telescoping the boom in/out; hoisting up/down; swinging the boom left/right and extending/retracting outrigger beams and/or stabilizers.

In addition to the pre-operational inspection, the operator should monitor the crane during operation for any potential deficiencies or hazards.

Operational Aids

Prior to each use, particular care should be taken to check all operational aids such as load weight indicators, anti-two-block devices, load moment indicators, boom angle indicators, boom length indicators, etc. This must be done in accordance with the manufacturer's written instructions.

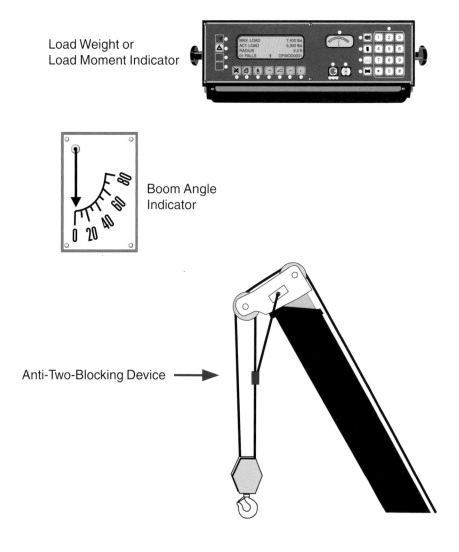

Load Weight or
Load Moment Indicator

Boom Angle
Indicator

Anti-Two-Blocking Device

Cranes Not in Regular Service

Before being used, cranes which have been idle for one to six months must be given a *frequent* inspection by a qualified person.

Standby cranes must also be inspected to this level at least every six months.

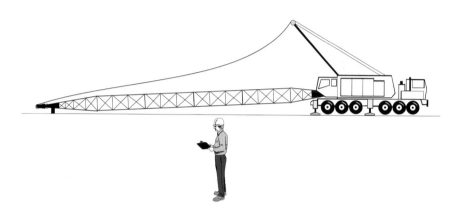

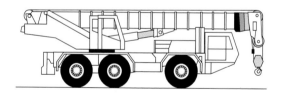

A periodic inspection must be performed on cranes idle for longer than six months.

Contents

Site Preparation

Setting the crane up properly is one of the most important aspects of a crane operation. In fact, a study of accidents reveals that over 50% of accidents are caused because the crane is set up improperly.

The person responsible for the crane operation must prepare the working area for the crane. This includes:

- Access roads, ground or supporting surfaces will support the crane and loads.
- Room to assemble and disassemble the crane.
- The crane barricaded to prevent entry of unauthorized personnel and the public.
- Power lines de-energized or clearances from power lines maintained.
- Blocking or mats available.
- Information such as load weights provided to operators and riggers.

Preparing the Work Area

Grade if necessary to ensure a firm and reasonably level site for the crane to travel and operate safely. Some site conditions may even require that the soil be compacted.

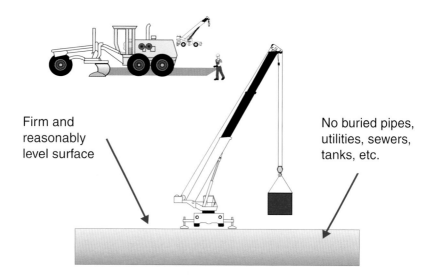

Firm and reasonably level surface

No buried pipes, utilities, sewers, tanks, etc.

If the crane is set up on a structure, the person responsible for the lift must ensure that the structure will support the crane and loads lifted.

Positioning the Crane

Contact with power lines is the leading cause of crane related fatalities. If possible, position the crane at least a boom's length away from the prohibited zone. See chapter "Working Cranes Around Power Lines" for more information.

PROHIBITED ZONE

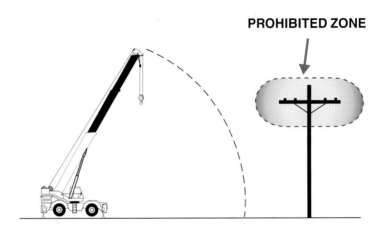

Try to position the crane so that the load is initially lifted from the least stable area of the crane and swung to the most stable area. If tipping should occur, it will be when the load is first lifted, rather than later in the lift should the load be higher off the ground.

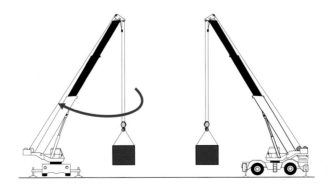

In the case of most rough terrain cranes, this would mean lifting the load over the side (least stable area) and swinging to the front (most stable area).

Maintaining Clearances

To prevent personnel from being struck or crushed, position the crane so that a minimum of two feet is maintained between the counterweight and any obstacle.

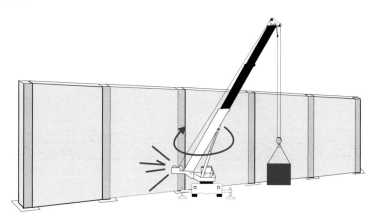

To further minimize the risk of personnel being injured, barricades should be erected around the swing radius of the outer most part of the crane. Unauthorized personnel must not be allowed inside this area.

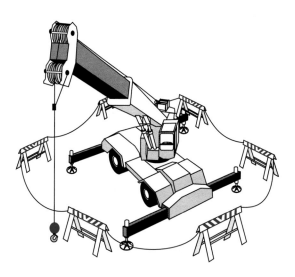

Maintaining Clearances

Allow enough room to assemble and disassemble the crane. In particular, ample space must be available for the erection and disassembly of lattice booms as well as manual sections, boom extensions and jibs.

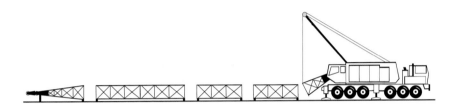

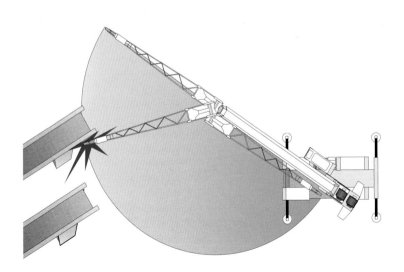

Dealing With Unstable Ground

Cranes should be set up a safe distance from buildings under construction, since the surrounding ground is often infill and a poor support base for the crane.

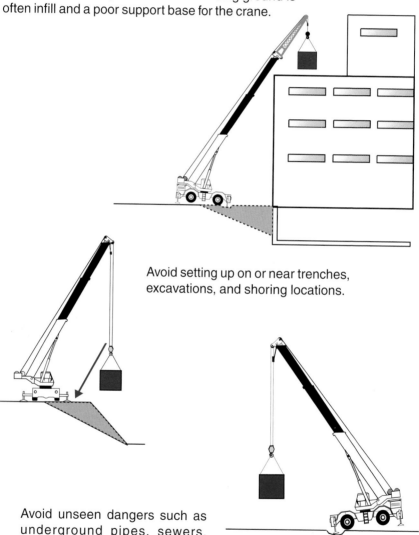

Avoid setting up on or near trenches, excavations, and shoring locations.

Avoid unseen dangers such as underground pipes, sewers, tanks, etc. Vibration, crane weight, and load weight could cause them to collapse.

Setup on Soft Surfaces

The combined weight of the crane and load must never exceed the bearing pressure of the surface where the crane is set up. On soft surfaces a working base formed by timber mats may have to be used to support the crane and loads.

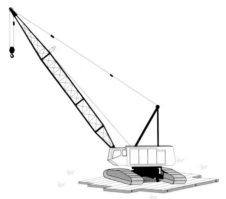

Lifting on outriggers, especially over just one, can create ground bearing pressures that are very high. Floats that are inadequately supported can cause the supporting surface to give way, resulting in the crane turning over.

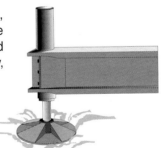

To prevent the crane from sinking or settling, timber or steel mats must be used when set up on surfaces such as soft ground, asphalt or backfilled material.

Use of Outriggers

Since the use of outriggers provide the crane with more stability than tires, outriggers should be used regardless of the weight of the load. Whenever possible they should be fully extended, or extended as specified by the manufacturer.

Each outrigger must be visible to the operator or to a signal person during extending or retracting.

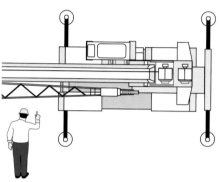

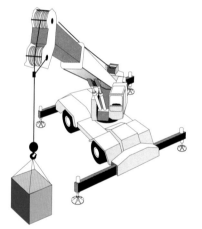

For the "on-outriggers" capacity chart to apply, all outriggers must be fully extended. Whenever one or more outrigger beams are left partially extended the "on-rubber" capacities must be used.

When loads are handled in an area where an outrigger is not fully extended, capacity loss can be over 50%.

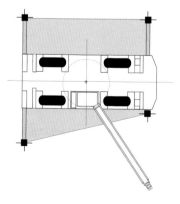

Use of Outriggers

The crane must be set up with all tires relieved of the crane's weight. Even better is to have the tires off the ground or supporting surface. This ensures that the tipping fulcrum is stabilized and the undercarriage is utilized to maximize counterweight.

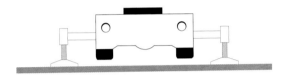

Before lifting with partially extended outrigger beams, make sure the manufacturer allows this practice. A partially extended outrigger beam can cause stress in the wrong areas, resulting in damage to the outrigger box and beam.

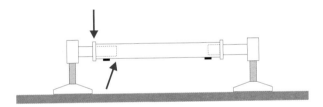

Some crane manufacturers provide capacity charts for lifting with partially extended outrigger beams. When doing so, instructions must be carefully followed, including pinning or securing beams if required.

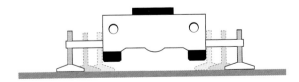

Use of Outriggers

Ensure that keepers are properly installed, otherwise the pad and stabilizer could separate during crane operations.

If stabilizers are equipped with screw down or pin locks, make sure these are engaged.

Use of Blocking

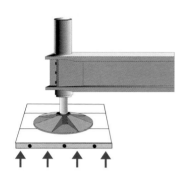

Blocking allows the weight of the crane and load to be distributed over a greater surface area. Blocking should be used under all outrigger floats.

Ensure blocking is level and that the outrigger pad is at 90° to the stabilizer.

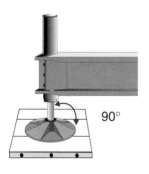

90°

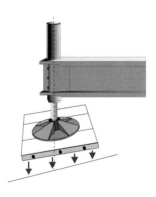

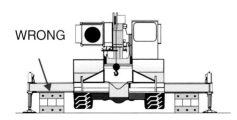

WRONG

Blocking must always be placed under the outrigger pads, never under the outrigger beams. Beams are not designed to take such loads, and the crane could become much less stable.

Use of Blocking

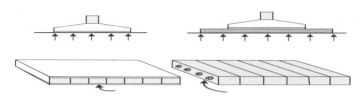

Without blocking,
ground pressure is higher.

With blocking,
ground pressure is reduced.

Use blocking that has the strength to prevent crushing, bending, or shear failure. Blocking must also be strong enough to span soft spots in the ground and strong enough to support the crane and load weight.

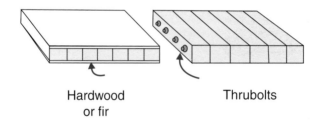

Hardwood
or fir

Thrubolts

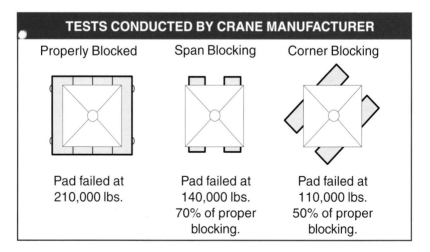

TESTS CONDUCTED BY CRANE MANUFACTURER

Properly Blocked	Span Blocking	Corner Blocking
Pad failed at 210,000 lbs.	Pad failed at 140,000 lbs. 70% of proper blocking.	Pad failed at 110,000 lbs. 50% of proper blocking.

Use of Blocking

As a general rule, blocking should be at least three times the surface area of the pad. Blocking must also completely support the float and transmit the load to the supporting surface.

Bolt blocking together to prevent it from separating.

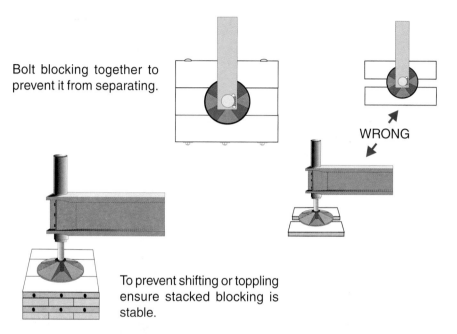

WRONG

To prevent shifting or toppling ensure stacked blocking is stable.

Blocking under the ends of the crawlers will improve crawler crane stability by keeping the tipping axis at the sprocket (or idler) and not at the roller. Crane must not be traveled off blocking with a load suspended.

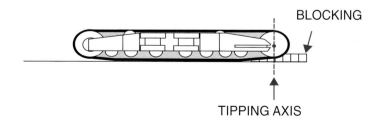

BLOCKING

TIPPING AXIS

Lifting On Rubber

Unless it is absolutely necessary, lifting while set up on tires should be avoided. Capacities are much less compared to lifting when on outriggers and the crane cannot be leveled when lifting on rubber.

However, when operations on rubber are required, the following areas, including the manufacturer's instructions, must be carefully observed:

- Carrier brakes are applied.
- Axle lockouts function properly.
- Working area is firm and level.
- Tires are in good condition and correct type and size.
- Tire pressure is per manufacturer's specifications.
- Wheels are in line with the carrier.
- Do not lift with jibs, extensions, or manual sections.
- Allow for tire and boom deflection.

BOOM DEFLECTION

TIRE DEFLECTION

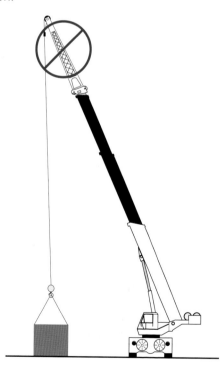

Leveling the Crane

Load chart capacities are based on the crane being level within 1%. Accurate leveling of the crane is therefore essential for safe lifting.

Use a bubble or target level for initial leveling, but don't rely on either for precision or final leveling.

Use a carpenter's level placed under or close to the boom foot for an accurate measure of level. Level the rear of the crane (front for rough terrain). Swing the boom 90° to the side and check for level again. Also check for level regularly during crane operation.

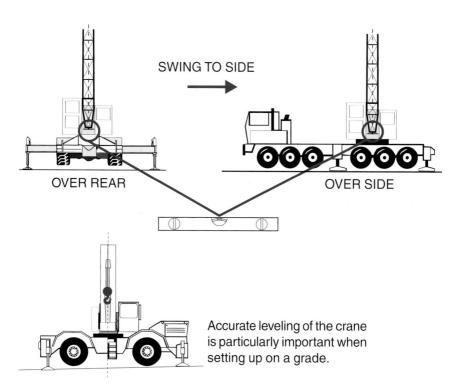

SWING TO SIDE

OVER REAR   OVER SIDE

Accurate leveling of the crane is particularly important when setting up on a grade.

Leveling the Crane

A similar two-part procedure (over front and over side) can also be used to check level using the hoist line as a plumb bob.

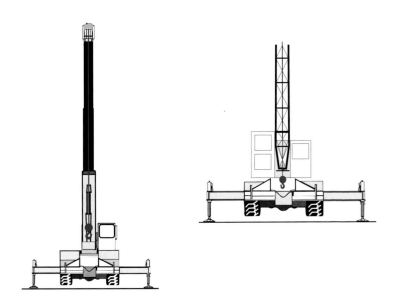

This procedure cannot be used to level a crawler crane or a crane working on rubber; they are only as level as the surface they are sitting on.

POSSIBLE CAPACITY LOSS DUE TO BEING OUT OF LEVEL

Boom Length and Radius	Capacity Lost When Out of Level		
	1°	2°	3°
Short Boom, Minimum Radius	10%	20%	30%
Short Boom, Maximum Radius	8%	15%	20%
Long Boom, Minimum Radius	30%	41%	50%
Long Boom, Maximum Radius	5%	10%	15%

This table shows the capacity loss that is possible for a specific lattice boom crane. Information like this can be obtained from all crane manufacturers.

Leveling the Crane

Particular care must be taken to ensure the supporting surface under crawler cranes and cranes operating "on rubber" is level before making the lift.

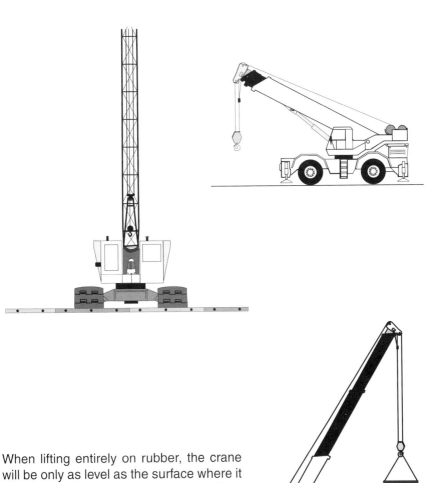

When lifting entirely on rubber, the crane will be only as level as the surface where it is sitting. Make sure the ground or surface is reasonably level before making the lift.

Contents

Pre-Lift Meeting

Before any work is begun on a site containing power lines, the person responsible for the operation must contact the utility company. It may be possible for the lines to be temporarily diverted around the job site, or to be de-energized and visibly grounded at the place of work.

Alternatively, insulating barriers may be erected to prevent physical contact between the crane, load, and lines. However, if none of these options are feasible, the following procedures must be observed.

An on-site meeting must be held between the person responsible for the job, the owner of the power lines (or a designated representative of the electrical utility), the operator, and all personnel involved in the lift – including any signal persons, riggers, etc.– to establish the procedures to be followed.

Everyone must understand these procedures, as well as the hazards of working near energized power lines, and how they can be avoided.

Maintaining Clearances

Maintain your distance. No part of the crane or load must ever enter the *"prohibited zone"* around a live power line. This zone must be enlarged as the kV increases (see table below.) Certain environmental conditions, such as fog, smoke or precipitation, may also require this distance to be increased.

Required clearance for operations near high voltage power lines:	
to 50 kV	10 ft. (3.05m)
over 50 to 200 kV	15 ft. (4.60m)
over 200 to 350 kV	20 ft. (6.10m)
over 350 to 500 kV	25 ft. (7.62m)
over 500 to 700 kV	35 ft. (10.67m)
over 750 to 1000 kV	45 ft. (13.72m)

kV = kilovolt (1000 volts) a unit of electrical potential difference.

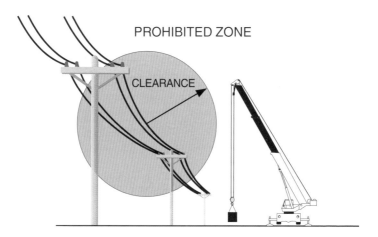

PROHIBITED ZONE

CLEARANCE

Consider erecting guard structures around the power lines, or use flags, balls or other highly visible devices to provide a constant reminder to all personnel.

Set the crane up as far as practically possible from the prohibited zone.

Use of a Signal Person

From the cab, it is difficult for the operator to judge distances accurately. Therefore, any time the crane is working within a boom's length of the prohibited zone, a *signal person* must be appointed. The signal person's **sole responsibility** is to warn the operator whenever any part of the crane or load approaches the *prohibited zone*.

No one may touch any part of the crane or the load, unless the signal person has indicated it is safe to do so.

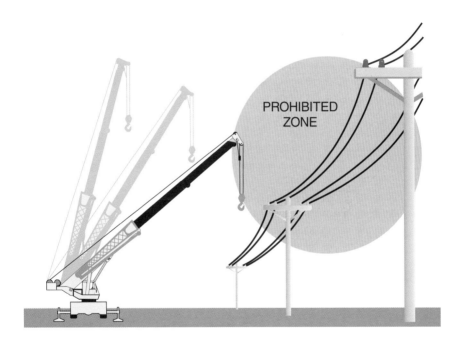

Consider using radio communications to ensure the signal person and operator remain in constant contact. Establish a procedure of voice commands, and make sure everyone follows them.

Wind and Tag Lines

Use caution near long spans as lines can move even in light wind. Add this distance to the minimum clearance required. (See clearances *p.108*). In certain environmental conditions, such as fog, smoke, rain or snow, you may also need to increase this clearance.

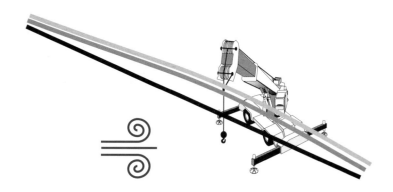

Whether hoisting, booming, swinging, or traveling, the crane must be operated *slowly* and with *extreme caution*.

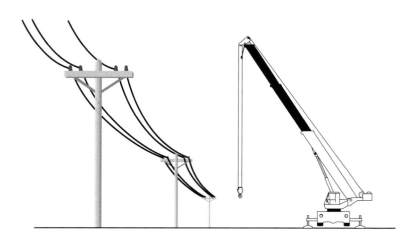

Working Above and Below Lines

Avoid operating cranes or handling materials above power lines.

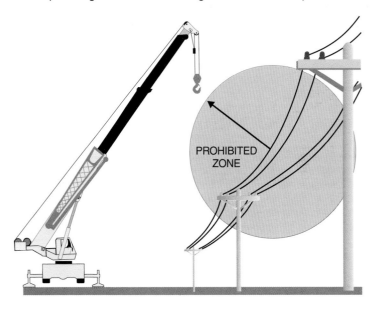

PROHIBITED ZONE

Because of unsafe situations that can be created, it is recommended that materials not be stored under power lines. Cranes must never be used to handle such materials if any combination of boom, load, load line, or machine component is capable of entering the prohibited zone.

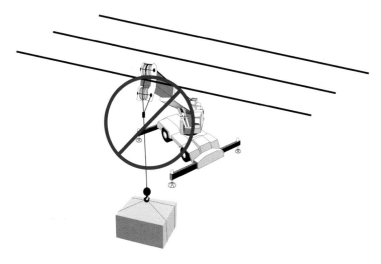

Warning Signs and Devices

Install durable signs at the operator's station and on the outside of the crane warning of the dangers of electrocution.

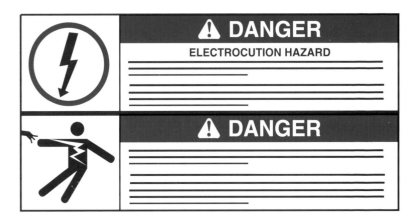

Consider the use of synthetic slings, which could reduce the effects of electrical contact. Some cranes may be equipped with devices such as insulated links (right), boom guards, and proximity warning devices.

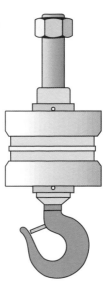

However, the ANSI/ASME B30.5 standard states that these devices have limitations and that, if they are used, all restrictions regarding operating near power lines still apply. Because of the lethal nature involved and potential of false security, make sure all personnel understand these limitations and that all instructions and testing requirements by the device manufacturer are followed.

Restricting the Work Area

When working around power lines, restrict the working area to essential personnel. A good way to accomplish this is by using barricades.

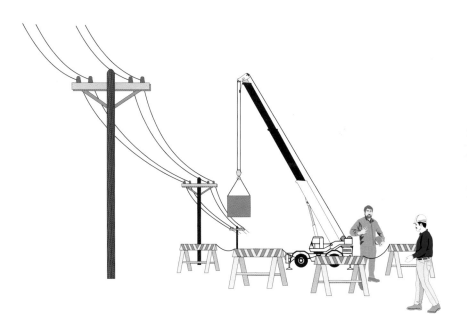

Consider erecting guard structures around the power lines, or use flags, balls or other highly visible devices to improve visibility and aid in location of the prohibited zone to all personnel.

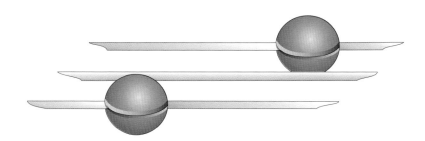

Driving Under Power Lines

If the crane has to be repositioned between lifts, and the route requires traveling the crane underneath power lines, use extreme caution.

Know the required clearances for operation in transit near power lines with no load, and boom or mast lowered. Remember that the motion of the crane over rough ground can cause the boom to bounce; this distance must be taken into account.

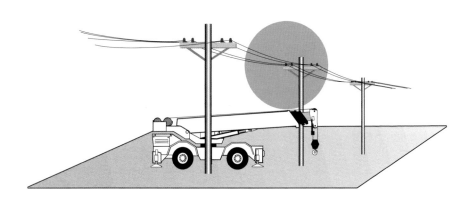

Clearances must be maintained when
in transit near high voltage power lines:

Required clearances for operation in transit near high voltage power lines:	
to 0.75 kV	4 ft. (1.22m)
0.75 to 50 kV	6 ft. (1.83m)
over 50 to 345 kV	10 ft. (3.05m)
over 345 to 750 kV	16 ft. (4.87m)
over 750 to 1000 kV	20 ft. (6.10m)

Response to Contact

If your crane does come into contact with a power line, **REMAIN IN THE CAB**.

DO NOT PANIC. You should be safe so long as you stay at a constant voltage within the cab.

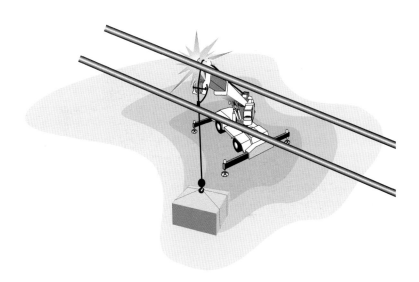

Instruct all personnel to **KEEP AWAY** from the crane and load, including anything attached to it such as hoist lines and tag lines. The ground around the crane will also likely have been energized.

Response to Contact

Try and disengage the crane from contact, and move it an appropriate distance from the power line.

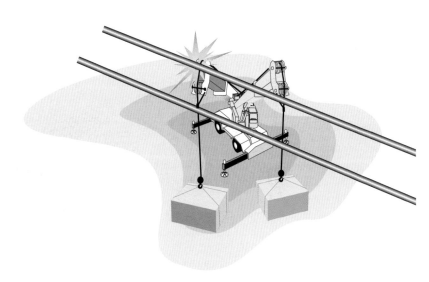

If contact cannot be broken, the operator should remain in the cab until the lines can be de-energized.

Emergency Evacuation

If it is necessary to leave the crane cab before the lines are de-energized (such as in the case of fire), do <u>not</u> climb down – **JUMP!** Do not make contact with the crane and ground at the same time; Doing so could be fatal.

Do not run or take long strides – electric current goes to ground in gradients and the voltage differential between gradients can kill. Instead, slowly shuffle away or take short jumps with feet firmly together.

Once safely clear of the crane, seek medical attention.

Post-Contact Procedure

Report the contact to the responsible authorities, such as the utility or owner of the power lines. The power lines themselves may need repair.

Also, thoroughly inspect the crane to ensure it was not damaged as a result of the contact. If in doubt about possible damage, contact the manufacturer. Wire rope, in particular, may need to be replaced.

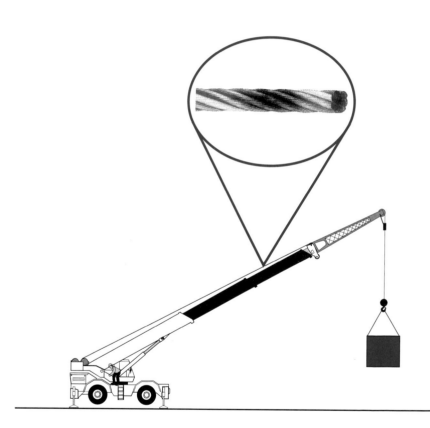

Other Electrical Sources

A crane can also become electrically charged when working close to transmission towers.

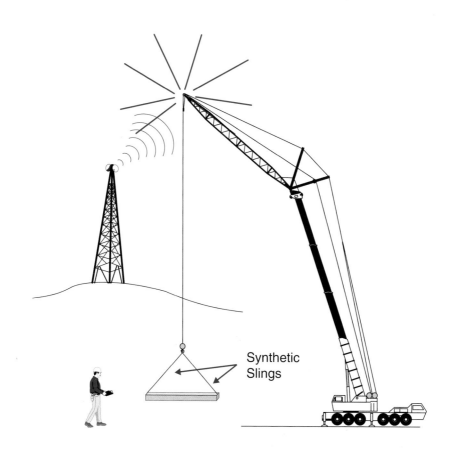

Synthetic Slings

Riggers should take care when handling suspended loads, and the use of synthetic slings is recommended.

Crane operators should wear rubber gloves when getting on and off the crane.

119

Contents

Getting On and Off the Crane

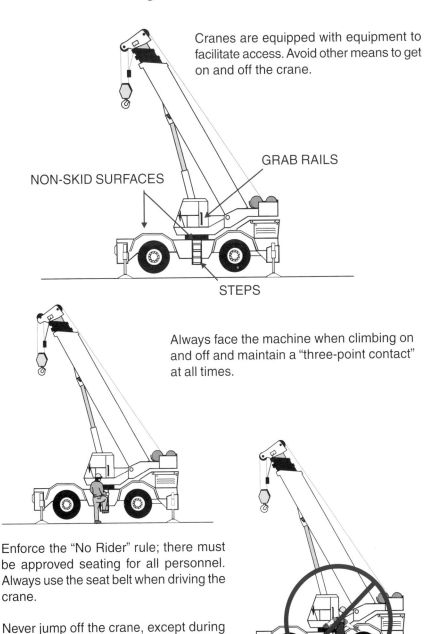

Cranes are equipped with equipment to facilitate access. Avoid other means to get on and off the crane.

GRAB RAILS

NON-SKID SURFACES

STEPS

Always face the machine when climbing on and off and maintain a "three-point contact" at all times.

Enforce the "No Rider" rule; there must be approved seating for all personnel. Always use the seat belt when driving the crane.

Never jump off the crane, except during emergencies.

Extending Boom Sections

By using sequencing valves, most modern telescoping boom cranes automatically extend the boom sections *simultaneously* by the use of a single telescoping lever. Cranes having multiple telescoping levers normally require that the operators themselves extend the boom sections equally.

CORRECT EXTENSION

On cranes where the telescoping sequence is not automatic, operators must consult the operator's manual and ensure they are familiar with the correct procedure.

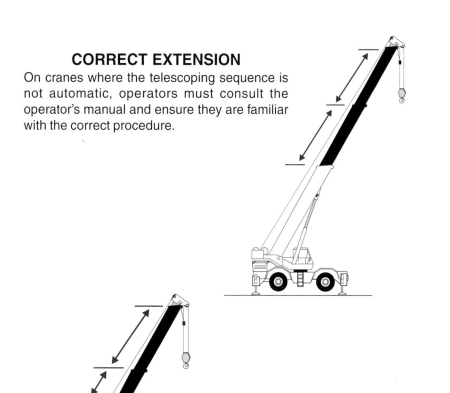

INCORRECT EXTENSION

Failure to extend boom sections according to the manufacturer's requirements may result in the boom bending or instability of the crane.

Hoisting Procedures

Before hoisting ensure:

- The hoist rope is not kinked.
- Multiple part lines are not twisted.
- The hook is above the load so as to prevent swinging when it is lifted.
- The rope is seated properly on the drum and in the sheaves (especially important if there has been a slack rope condition).
- Weather conditions are acceptable.

During lifting, avoid sudden starts or stops; these can cause dynamic loading and seriously damage the crane. Ensure no part of the crane or load can contact any obstruction.

NEVER use a crane to drag a load sideways. Side loading of booms is permitted on freely suspended loads ONLY.

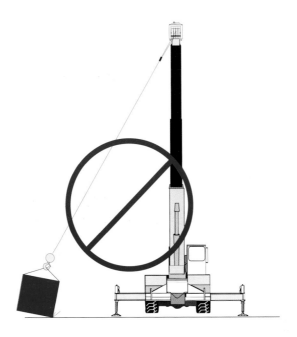

Hoisting Procedures

Before and during hoisting operations, the person directing the lift must ensure:

- The crane is level and, where necessary, blocked.
- The load is secured and balanced in the sling or lifting device.
- Both the lift and swing path are clear of obstructions and people.
- All persons are clear of the swing radius.

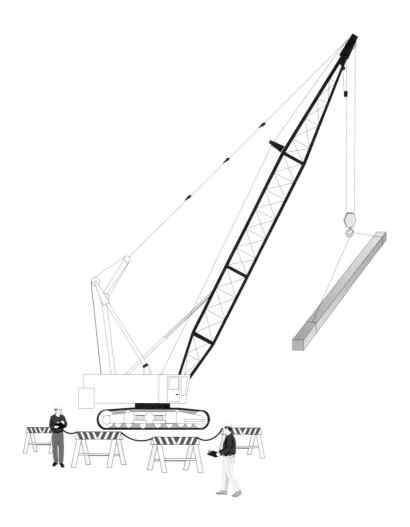

Avoiding Distractions

The operator should never be distracted while operating the crane. His safety and the safety of other workers depend on his constant attention on the job at hand.

The operator must obey only the signals given by the appointed signal person or the person directing the lift . . . with the sole exception of the stop signal, which must be obeyed regardless of who gives it. Hand signals should be in accordance with those specified in ANSI/ASME B30.5.

Operators must take signals from only one designated signal person - never both simultaneously.

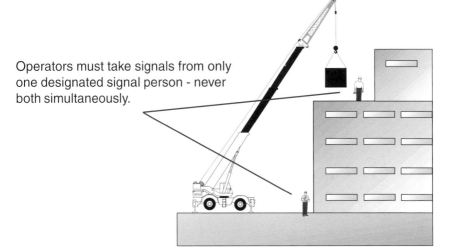

Dangers of Tipping

As a crane starts to tip, the load radius increases. As the center of gravity of the load moves away from the tipping axis, the center of gravity of the crane moves toward the tipping axis. This accelerates the rate of tipping and may leave the operator powerless to remedy the situation other than by dropping the load.

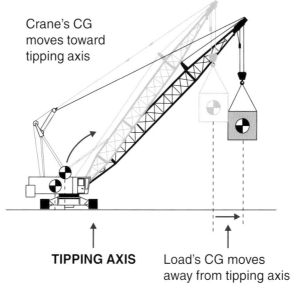

Crane's CG moves toward tipping axis

TIPPING AXIS   Load's CG moves away from tipping axis

Tipping on telescoping boom cranes may be even more rapid because of the greater weight of the boom.

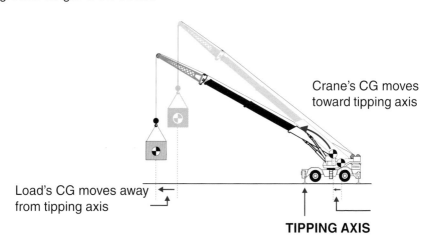

Crane's CG moves toward tipping axis

Load's CG moves away from tipping axis

TIPPING AXIS

Dangers of Tipping

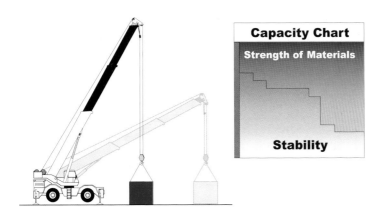

In the structural area of the load chart, crane capacities are based on the strength of its components. This means a crane may fail structurally before it tips ... so **never** use signs of tipping (i.e. "operating by the seat of the pants") as an indication of a crane's ability to lift. When a crane starts to tip it is already overloaded and may incur structural damage.

Never use signs of tipping as an indication of a crane's ability to lift.

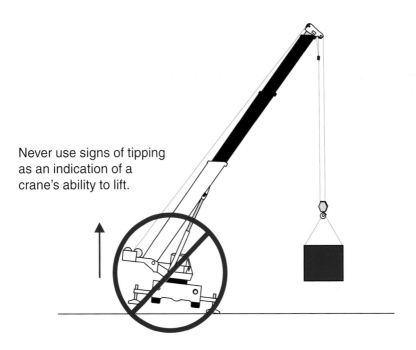

Two-Blocking

Two-blocking occurs when the hook block or headache ball makes contact with the sheaves at the main boom head, extension or jib tip. This can break the hoist rope and cause the hook block or ball to fall.

Two-blocking occurs most commonly on telescoping boom cranes as a result of over-hoisting or telescoping the boom without letting out the hoist line.

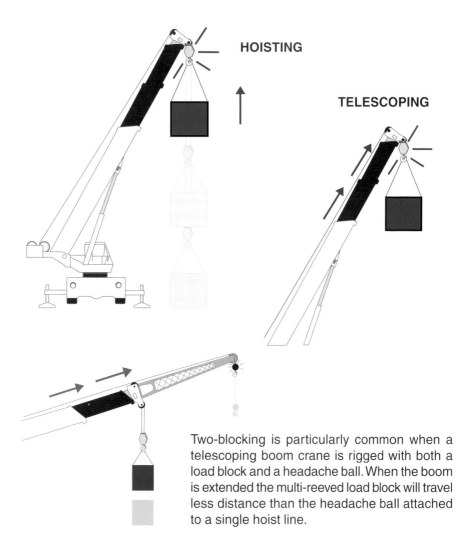

HOISTING

TELESCOPING

Two-blocking is particularly common when a telescoping boom crane is rigged with both a load block and a headache ball. When the boom is extended the multi-reeved load block will travel less distance than the headache ball attached to a single hoist line.

Two-Blocking

BOOMING DOWN

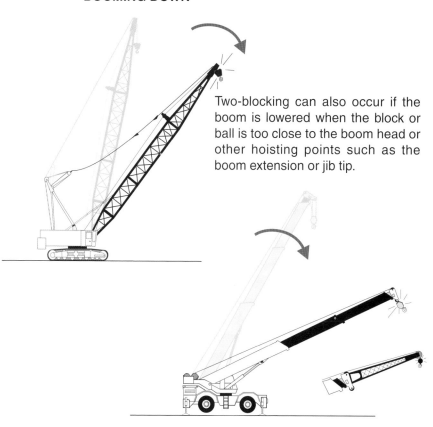

Two-blocking can also occur if the boom is lowered when the block or ball is too close to the boom head or other hoisting points such as the boom extension or jib tip.

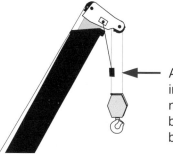

An anti-two-block device can stop an impending two-block condition and, if necessary, prevent further hoisting, boom extension and lowering of the boom.

Backward Collapse of Boom

Contrary to popular belief, boom stops are not designed to stop the boom from collapsing backwards. Ensure the boom hoist deactivates before the boom contacts the boom stops.

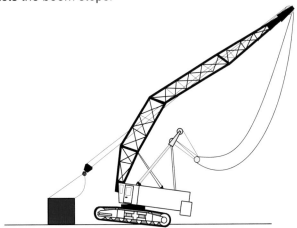

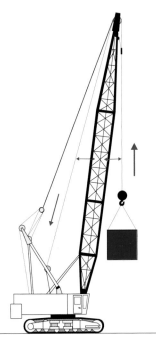

The boom is at most risk of going over backwards when the angle between the load line and the boom is the same or close to the angle between the hoist line and the boom. The heavier the load the greater the risk.

Risk is increased when the rear hoist drum is being used and the crane is rigged with a single part of line.

Backward Collapse of Boom

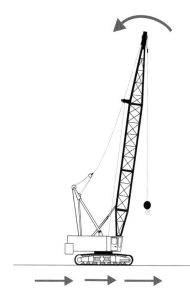

Backward collapse of the boom can occur when traveling the crane with the boom at a high angle. In particular, sudden stops and starts should be avoided.

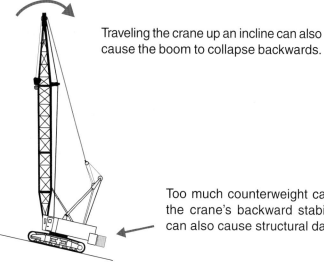

Traveling the crane up an incline can also cause the boom to collapse backwards.

Too much counterweight can affect the crane's backward stability and can also cause structural damage.

Backward Collapse of Boom

Swinging the crane from the downhill side to the uphill side can cause the boom to collapse backwards.

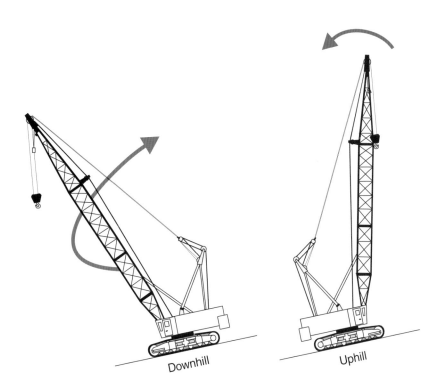

Downhill   Uphill

Cranes should be operated in a level position. Operating while unlevel not only reduces crane capacity, but causes the load radius to change as the crane rotates.

Backward Collapse of Boom

Two-blocking can cause the boom to be pulled backwards over the crane. This is even more likely to occur when the boom is operating at a high angle.

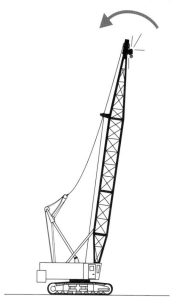

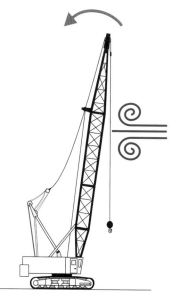

Operating the crane in high winds can have the same effect, especially when using long booms at high boom angles.

Backward Collapse of Boom

Backwards collapse of the boom can occur when tightening a load line that is connected back to the boom foot.

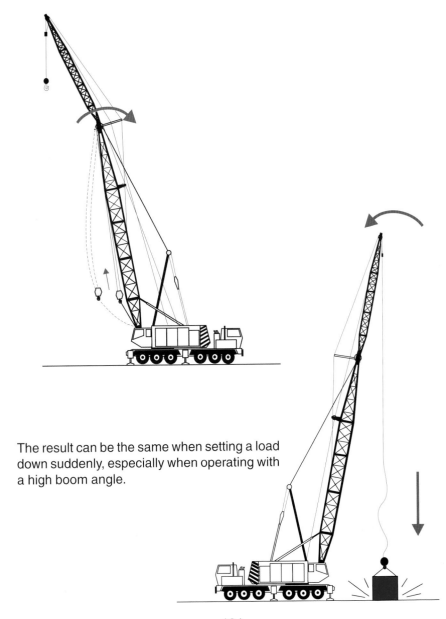

The result can be the same when setting a load down suddenly, especially when operating with a high boom angle.

Installing Wire Rope

The wire rope on all load and boom hoist drums must be wound tightly and evenly at all times. Severe damage can occur from improperly spooled drums. Among the leading causes of poor drum winding and slack rope is the rope incorrectly installed on the drum.

Use these simple methods to determine the correct installation of wire rope.

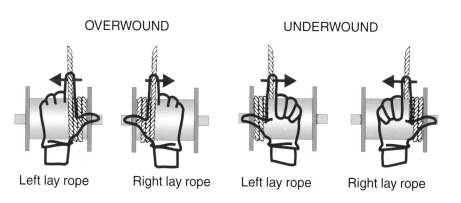

OVERWOUND		UNDERWOUND	
Left lay rope	Right lay rope	Left lay rope	Right lay rope

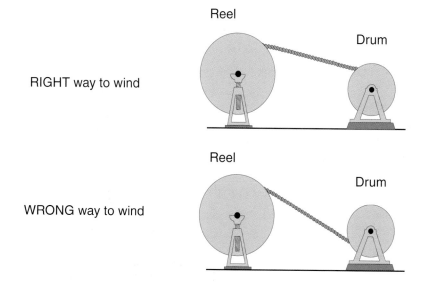

RIGHT way to wind

WRONG way to wind

Avoiding Uneven Winding

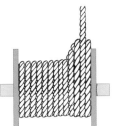

Uneven winding can occur when the fleet angle is incorrect. This can result in the load dropping or wire rope damage.

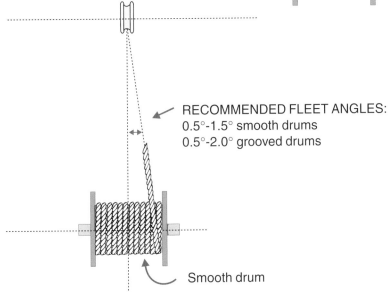

RECOMMENDED FLEET ANGLES:
0.5°-1.5° smooth drums
0.5°-2.0° grooved drums

Smooth drum

Always maintain at least 2 rope wraps on drum. Some authorities require 3 wraps.

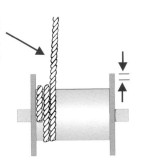

Flange is to be at least 1/2" above top layer of rope.

Avoiding Uneven Winding

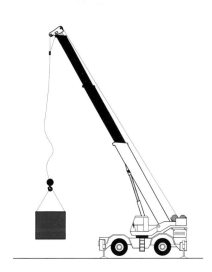

Uneven winding can occur if the load is suddenly stopped as it is being hoisted, causing the load to bounce.

Uneven winding can occur if the rope is too stiff, or the sheaves are poorly lubricated.

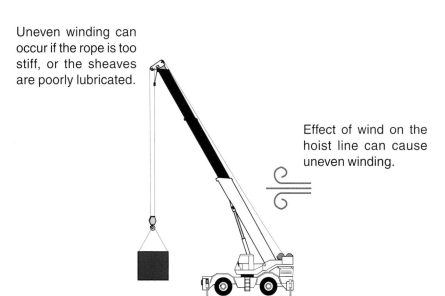

Effect of wind on the hoist line can cause uneven winding.

Avoiding Slack Wire Rope

Uneven winding and/or slack wire rope can also be caused if:

...the headache ball is too light. As the boom is raised the weight of the headache ball may be insufficient to counter the force of the winch, causing the rope to become slack or even to be pushed off the drum.

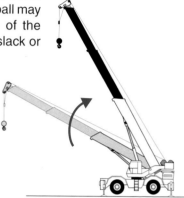

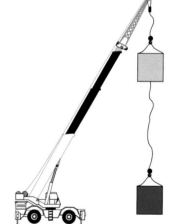

...there is a sudden change in rope tension caused by setting a load down too fast, or suddenly releasing a load.

Overlowering the hoist line once the load is placed can have the same effect.

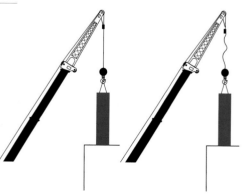

Avoiding Boom Damage

Never allow a crane boom to hit any structure. This may dent, bow, or bend the boom chords, or even buckle the boom.

Even minor damage can seriously weaken a boom and even result in total collapse.

If the boom, mast, or gantry makes contact with a structure, **do not** lower the boom until a qualified person assesses the damage and indicates it is safe to do so. An assist crane may have to be used to help lower a damaged boom.

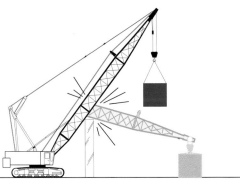

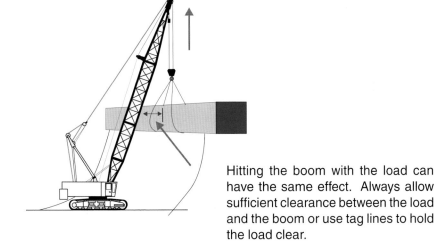

Hitting the boom with the load can have the same effect. Always allow sufficient clearance between the load and the boom or use tag lines to hold the load clear.

Tilt-Up Operations

Be sure the crane can safely handle and suspend loads that are lifted from a horizontal to a vertical position. As the load is lifted, the radius will increase and the crane's capacity will decrease.

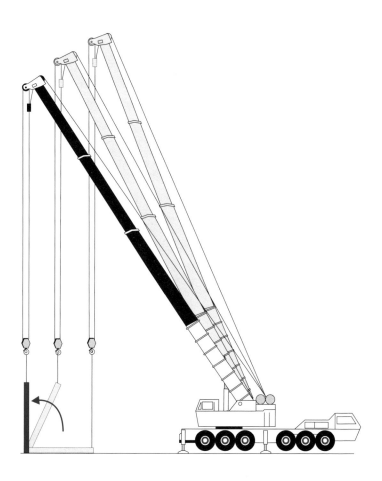

Multi-Crane Lifts

Using more than one crane to lift and place a load is a complex procedure. Doing this safely requires a high level of skill and must be meticulously planned, preferably by an engineer. Responsibility for the operation must be assumed by a qualified person who must analyze the operation and plan for all eventualities. The plan should include the following requirements:

- All personnel involved in the operation must be qualified to perform their duties.
- All personnel must be instructed in all phases of the operation.
- The ground or supporting surface must be capable of supporting the weight of both the crane and its load.
- All cranes must be set up on mats or blocking.
- All cranes must be level and positioned properly.
- The load weight imposed on each crane must be accurately determined.
- Each crane's net capacity must be determined exactly.
- All movement must be made in a slow and controlled manner.
- Consider reducing each crane's net capacity 25%.

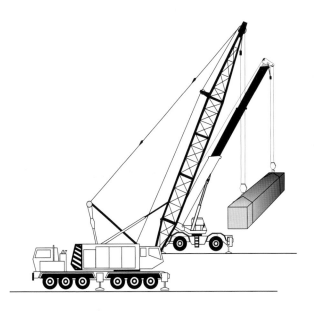

Traveling the Crane

Before traveling on site *with a load*, be sure this is *allowed* by the manufacturer. If permitted, designate a person to be responsible for the operation. In addition to the manufacturer's instructions, areas that must be considered in pick-and-carry operations are:

- Reducing crane ratings.
- Length and position of boom.
- Load close to carrier and ground.
- Travel route and ground condition.
- Travel speed.
- Tire condition and pressure.
- Traveling on inclines (should be avoided).

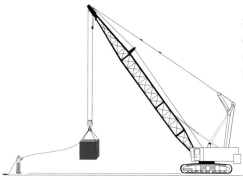

When required, use tag lines to control the load. Also ensure the boom is low enough to avoid the possibility of backward collapse.

Most manufacturers do not permit travel with a load on any pinned section, boom extension or jib.

Traveling the Crane

Before traveling up an incline, make sure the boom is lowered to prevent the boom from collapsing backwards.

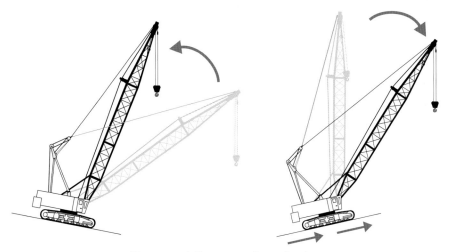

To prevent the crane from tipping, the boom should be raised before traveling down an incline.

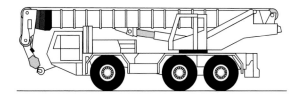

When traveling the crane from one job site to another, ensure:

- The boom is in line with the direction of motion.
- The swing brake is engaged (unless the boom is supported on a dolly to negotiate turns).
- The hook is secured to prevent it from swinging.

Directing the Operator

Directing the lift is one of the most important jobs in a crane operation. Since the signal person is in a sense operating the crane, the person responsible for the operation must only designate qualified persons to direct the operator. To be considered qualified, a signal person must have a basic understanding of crane operation and limitations and a thorough understanding of standard hand signals.

The signal person is responsible for keeping nonessential personnel out of the work area, and must not direct the load over personnel.

The signal person when required must be positioned where the operator, path of travel and location where the load will be placed can clearly be seen.

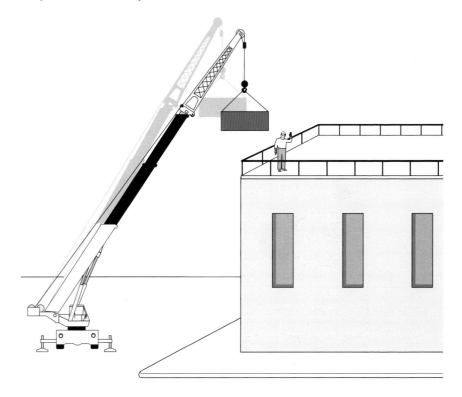

Directing the Operator

The crane operator must not respond to signals unless they are clearly understood. Except for the stop signal, which anyone can give, the operator must only respond to signals given by the designated signal person. Unless voice communication equipment is used, standard signals used to direct the operator must be those prescribed in ANSI/ASME B30.5.

STANDARD SIGNALS

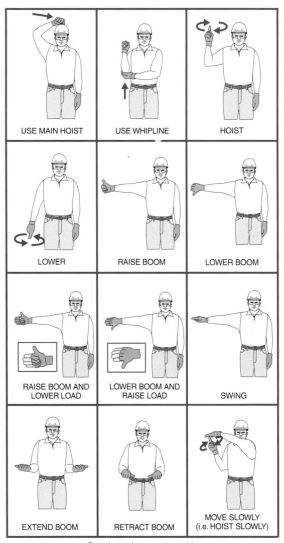

Continued on page 146

Directing the Operator

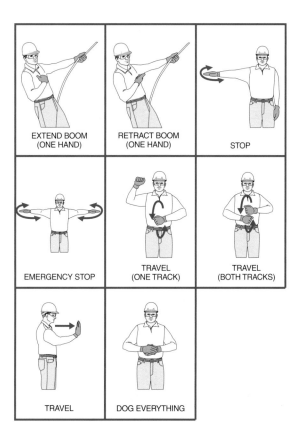

EXTEND BOOM (ONE HAND)	RETRACT BOOM (ONE HAND)	STOP
EMERGENCY STOP	TRAVEL (ONE TRACK)	TRAVEL (BOTH TRACKS)
TRAVEL	DOG EVERYTHING	

When operations or conditions exist which are not covered by standard signals, special signals may be used. However, they must be agreed upon before hand by the operator and signal person and must not conflict with standard signals.

Directing the Operator

When moving a carrier-mounted crane with two cabs, the following audible signals must be used.

Stop - One audible signal

Go ahead - Two audible signals

Back up - Three audible signals

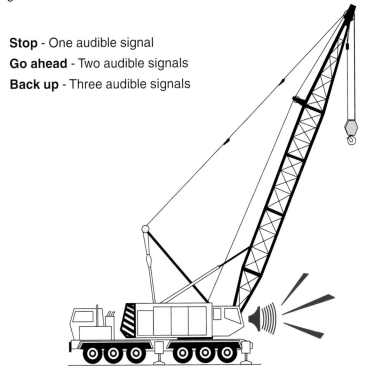

Leaving Crane Unattended

Before leaving the crane unattended, the operator must:

- Land any load, bucket, lifting magnet, or other device.
- Disengage master clutch.
- Place all control levers in neutral.
- Set all brakes and locking devices.
- Secure the crane against accidental travel.
- Shut off the engine.

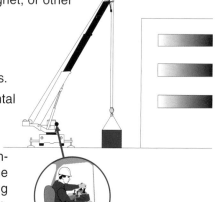

If the crane operation is frequently interrupted, the operator may leave the crane with the engine running so long as the crane remains in view and unauthorized entry is prevented.

If a load is held suspended for a period of time, the operator may leave the controls provided that the following precautions are established:

- Restrain boom hoist, telescoping, load, swing and outrigger functions.
- Provide notices, barricades or other precautions if necessary to ensure safety.

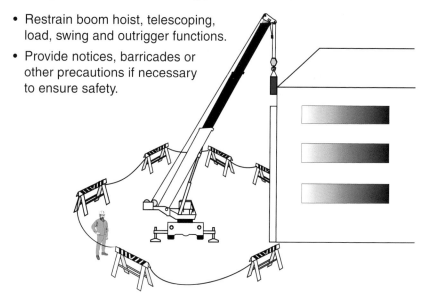

Power Failure

In the event of power failure, the operator should:

- Try to lower the load to the ground.
- Set all brakes and locking devices.
- Move all controls to the off or neutral position.

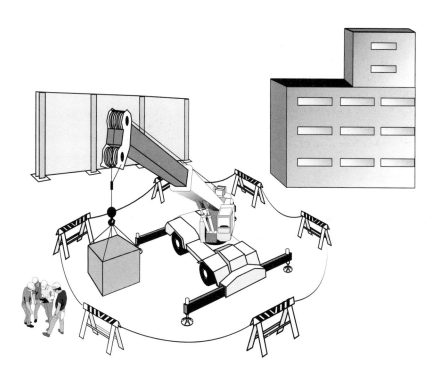

Crane Shutdown

When leaving cranes unattended overnight, or for longer periods of time, the following procedure should be followed:

- Set load on ground.
- Retract and/or lower the boom if possible.
- If the crane is set up on outriggers, leave it that way.
- Place control levers in neutral.
- Disengage master clutch (if applicable).
- Set all brakes and locking devices.
- Shut down engine.
- Lock all doors to prevent unauthorized entry.

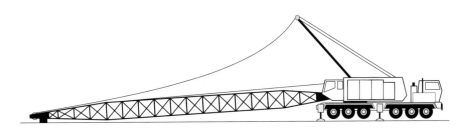

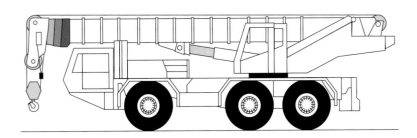

Contents

Pre-Lift Considerations

Before any hoisting of personnel by crane begins, the person responsible for the job must establish that there is no less hazardous way of performing the job.

To help ensure that proper procedures are followed, the person responsible for the task must hold a meeting with all personnel involved in the operation. This will include the crane operator, signal person(s) (if necessary for the lift), and the personnel to be lifted. The meeting must be held prior to the trial lift at each new work location, and each time employees are newly assigned to the operation.

A specially designed personnel platform, conforming to OSHA specifications, must be used. Riding the load, or the headache ball, or any other method is NOT permitted.

Platform Specifications

The platform must be designed by a qualified person and all welding performed by a qualified welder. In particular, each platform must have:

- a design factor of 5:1.
- a suspension system to minimize tipping.
- a guardrail and an inside grab rail.
- sufficient headroom for personnel to stand.
- no rough edges which might snag personnel.
- permanent indication of its weight and rated capacity.

Overhead protection must be provided where personnel could be exposed to falling objects (hard hats, though required, are insufficient by themselves).

Each side of the platform must be enclosed to mid-rail.

Access gates (if fitted) must swing inwards only and be prevented from opening accidentally.

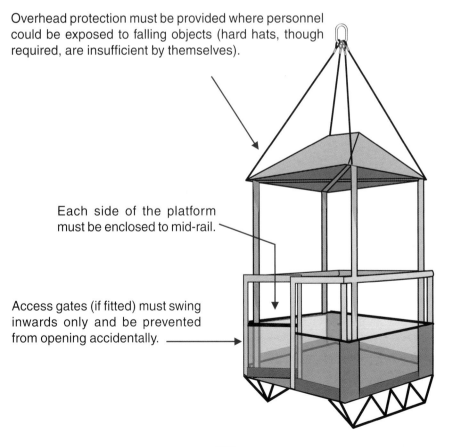

Selection of the Crane

In addition to meeting standard criteria for inspection and maintenance, cranes used for personnel lifting must be derated by 50%, i.e. all capacities shown in the crane's load chart must be halved.

Cranes must be equipped with:

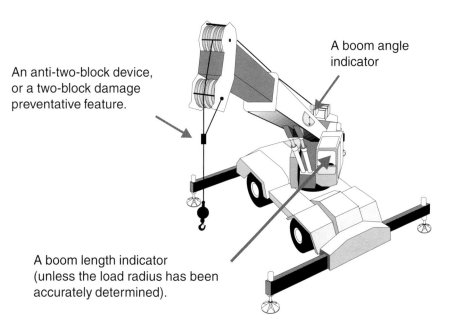

An anti-two-block device, or a two-block damage preventative feature.

A boom angle indicator

A boom length indicator (unless the load radius has been accurately determined).

The crane selected for the personnel hoisting operation must also be equipped with controlled load lowering. No free fall is permitted.

Load lines must be able to lift seven times the maximum intended load (i.e. a 7:1 design factor), or ten times for rotation resistant ropes (i.e. a 10:1 design factor).

Selection of Rigging

The rigging selected for personnel hoisting must not be used for any other purpose and should be kept apart from other rigging or clearly identified in some way. It must be capable of handling 5 times the maximum intended load (10 times for rotation resistant wire rope).

NOTICE
THIS RIGGING TO BE USED FOR PERSONNEL LIFTING ONLY. ALL OTHER USES PROHIBITED.

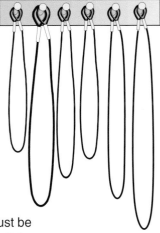

Wire rope bridles must be attached to a master link or shackle to ensure even load distribution.

All hooks must have lockable latches.

Eyes in wire rope slings must be fabricated with thimbles.

155

Trial and Test Lifts

The usual correct procedure for setting up the crane must be followed *(see chapter 5, **Crane Setup**)*. In particular, the crane must be on a firm surface and level to within 1% of grade.

Before any hoisting with a personnel platform begins, a proof test and a trial lift must be conducted.

A **PROOF TEST** must be conducted at each new job site to 125% of the platform's rated capacity. The load must be evenly distributed and held suspended for five minutes.

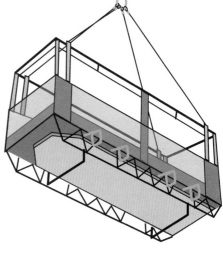

A **TRIAL LIFT** (which can be combined with the proof test) must be conducted for each location the platform will be lifted to. This trial lift must also be repeated each time the crane is moved to a new position.

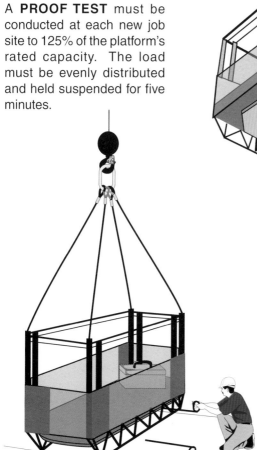

Final Pre-Lift Inspection

A visual inspection of the crane, rigging, and platform must be made by a competent person after the trial lift is completed.

A final visual inspection must be made by a competent person after the platform has been raised a few inches off the ground, but before hoisting of personnel begins.

In particular, checks must be made to ensure the hoist rope is:

- not twisted.
- free of kinks.
- centered over platform.
- properly seated on drums and in sheaves.

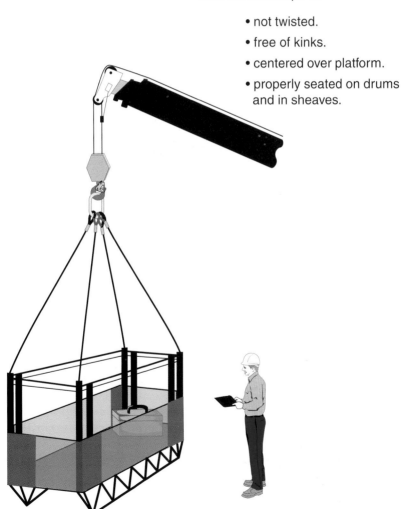

157

Preparing to Lift

Care must be taken to ensure the platform is loaded evenly and within its rated capacity. Only the personnel required to perform the job are permitted to be in the platform.

Unless the work is being done over water, all personnel must be tied off to the platform or load block/headache ball.

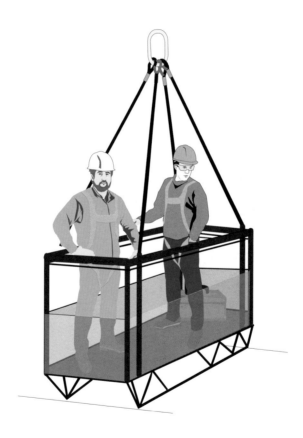

Necessary tools and materials are permitted in the platform as long as they are evenly distributed and properly secured. The platform must not to be used as a materials hoist, i.e. with tools and materials alone.

Lifting to the Work Location

Hoisting must be slow, controlled and cautious at all times. All parties must remain in constant visual or radio communication.

With the sole exception of the signal person, personnel in the platform must take care to keep all body parts inside during hoisting operations.

Tag lines must be used to help control the platform unless their use creates an unsafe condition.

At the Work Location

Once the platform has reached the work location, the crane operator must engage all brakes and remain at the controls.

Before any personnel are allowed to leave the platform, it must be landed on, or tied off to a structure, unless it would be unsafe to do so.

No lift may be made on the crane's other hoist line while personnel are in the platform.

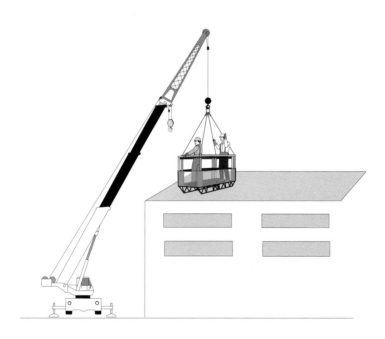

Lifting Over Water

When working over water, personnel in the platform need not be tied off, but they must be supplied with U.S. Coast Guard-approved life jackets.

In addition, ring buoys, each with 90 feet of line, must be provided at least every 200 feet around the work site.

At least one lifesaving skiff must be on hand for emergency rescue.

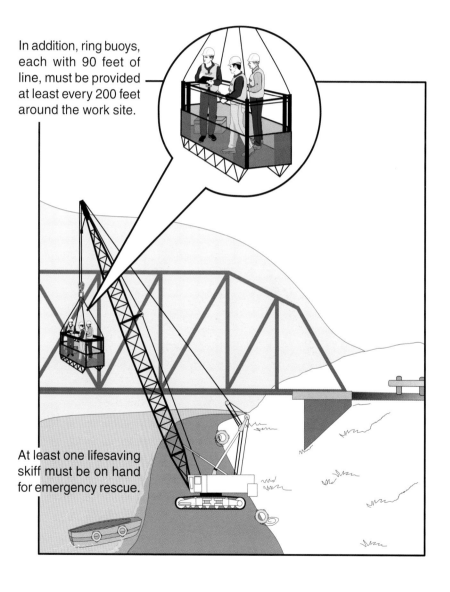

Traveling the Crane

Traveling the crane with personnel is prohibited by ANSI/ASME B30.5. However, OSHA allows this practice provided there is no less hazardous way to perform the work. Be sure to exercise extreme caution when traveling cranes with personnel.

The employer must implement the following procedures:

• Crane travel must be restricted to a fixed track or runway.
• Travel must be limited to the load radius of the boom used during the lift.
• The boom must remain in line with the direction of travel.

In addition to the proof test and trial lift required for stationary lifting, a trial run must be performed to test the route of travel before personnel enter the platform.

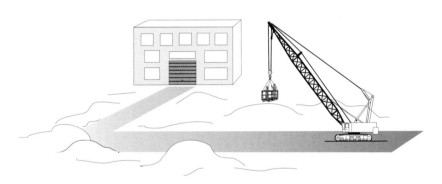

For rubber tire cranes, the on-rubber capacities must be used and <u>derated by 50%</u>. Special attention during the pre-lift inspection should be given to the condition of the crane's tires, including air pressure.

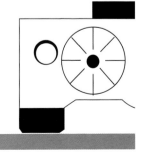

Contents

Wire Rope
6 x 19 and 6 x 37 Classification

Bright (Uncoated) • IWRC									
Nominal Diameter		Approximate Mass		Minimum Breaking Strength*					
				IPS**		EIPS**		EEIPS	
inches	mm	lb/ft	kg/m	tons	metric tons	tons	metric tons	tons	metric tons
3/8	9.5	0.26	0.39	6.56	5.95	7.55	6.85	8.3	7.54
7/16	11.5	0.35	0.52	8.89	8.07	10.2	9.25	11.2	10.18
1/2	13	0.46	0.68	11.5	10.4	13.3	12.1	14.6	13.3
9/16	14.5	0.59	0.88	14.5	13.2	16.8	15.2	18.5	16.7
5/8	16	0.72	1.07	17.9	16.2	20.6	18.7	22.7	20.6
3/4	19	1.04	1.55	25.6	23.2	29.4	26.7	32.4	29.4
7/8	22	1.42	2.11	34.6	31.4	39.8	36.1	43.8	39.7
1	26	1.85	2.75	44.9	40.7	51.7	46.9	56.9	51.6
1 1/8	29	2.34	3.48	56.5	51.3	65.0	59.0	71.5	64.9
1 1/4	32	2.89	4.30	69.4	63.0	79.9	72.5	87.9	79.8
1 3/8	35	3.50	5.21	83.5	75.7	96.0	87.1	106	95.8
1 1/2	38	4.16	6.19	98.9	89.7	114	103	125	113
1 5/8	42	4.88	7.26	115	104	132	120	146	132
1 3/4	45	5.67	8.44	133	121	153	139	169	153
1 7/8	48	6.5	9.67	152	138	174	158	192	174
2	52	7.39	11.0	172	156	198	180	217	198
2 1/8	54	8.35	12.4	192	174	221	200	243	220
2 1/4	57	9.36	13.9	215	195	247	224	272	246
2 3/8	60	10.4	15.5	239	217	274	249	301	274
2 1/2	64	11.6	17.3	262	238	302	274	332	301
2 5/8	67	12.8	19.0	288	261	331	300	364	330
2 3/4	70	14.0	20.8	314	285	361	327	397	360
2 7/8	74	15.3	22.8	341	309	392	356	431	392

* 1 ton = 2000 lbs. 1 metric ton = 2204 lbs.

* To convert to kilonewtons (kN), multiply tons (nominal strength) by 8.896;
  1 lb = 4.448 newtons (N).

** Available with galvanized wires at strengths 10% lower than listed, or at equivalent
  strengths on special request.

Wire Rope
8 x 19 Classification

Rotation Resistant • Bright (Uncoated) • IWRC

Nominal Diameter		Approximate Mass		Minimum Breaking Strength*			
				IPS**		EIPS**	
inches	mm	lb/ft	kg/m	tons	metric tons	tons	metric tons
1/2	13	0.47	0.70	10.1	9.16	11.6	10.5
9/16	14.5	0.60	0.89	12.8	11.6	14.7	13.3
5/8	16	0.73	1.09	15.7	14.2	18.1	16.4
3/4	19	1.06	1.58	22.5	20.4	25.9	23.5
7/8	22	1.44	2.14	30.5	27.7	35.0	31.8
1	26	1.88	2.80	39.6	35.9	45.5	41.3
1 1/8	29	2.39	3.56	49.8	45.2	57.3	51.7
1 1/4	32	2.94	4.37	61.3	55.6	70.5	64.0
1 3/8	35	3.56	5.30	73.8	67.0	84.9	77.0
1 1/2	38	4.24	6.31	87.3	79.2	100.0	90.7

* 1 ton = 2000 lbs. 1 metric ton = 2204 lbs.

* To convert to kilonewtons (kN), multiply tons (minimum breaking strength) by 8.896; 1 lb = 4.448 newtons (N).

** Available with galvanized wires at strengths 10% lower than listed, or at equivalent strengths on special request.

NOTE: The given strengths for 8 x 19 rotation resistant ropes are applicable only when a test is conducted on a new rope fixed at both ends. When the rope is in use, and one end is free to rotate, the nominal strength is reduced.

Wire Rope
19 x 7 Classification

Rotation Resistant • Bright (Uncoated)

Nominal Diameter		Approximate Mass		Minimum Breaking Strength*			
				IPS**		EIPS**	
inches	mm	lb/ft	kg/m	tons	metric tons	tons	metric tons
1/2	13	0.45	0.67	9.85	8.94	10.8	9.8
9/16	14.5	0.58	0.86	12.4	11.2	13.6	12.3
5/8	16	0.71	1.06	15.3	13.9	16.8	15.2
3/4	19	1.02	1.52	21.8	19.8	24.0	21.8
7/8	22	1.39	2.07	29.5	26.8	32.5	29.5
1	26	1.82	2.71	38.3	34.7	42.2	38.3
1 1/8	29	2.30	3.42	48.2	43.7	53.1	48.2
1 1/4	32	2.84	4.23	59.2	53.7	65.1	59.1
1 3/8	35	3.43	5.10	71.3	64.7	78.4	71.1
1 1/2	38	4.08	6.07	84.4	76.6	92.8	84.2

* 1 ton = 2000 lbs. 1 metric ton = 2204 lbs.

* To convert to kilonewtons (kN), multiply tons (minimum breaking strength) by 8.896; 1 lb = 4.448 newtons (N).

** Available with galvanized wires at strengths 10% lower than listed, or at equivalent strengths on special request.

NOTE: The given strengths for 8 x 19 rotation resistant ropes are applicable only when a test is conducted on a new rope fixed at both ends. When the rope is in use, and one end is free to rotate, the nominal strength is reduced.

Wire Rope
Compacted 6 x 19 and 6 x 37 Classification

Compacted Strand • Bright (Uncoated) • FC & IWRC

Nominal Dia.		Approximate Mass				Minimum Breaking Strength*			
inches	mm	lb/ft		kg/m		tons		metric tons	
		FC	IWRC	FC	IWRC	FC	IWRC	FC	IWRC
3/8	9.5	.26	.31	.39	.46	7.39	8.3	6.7	7.53
7/16	11.5	.35	.39	.52	.58	10.0	11.2	9.07	10.2
1/2	13	.46	.49	.68	.73	13.0	14.6	11.8	13.2
9/16	14.5	.57	.63	.85	.94	16.4	18.5	14.9	16.8
5/8	16	.71	.78	1.06	1.16	20.2	22.7	18.3	20.6
3/4	19	1.03	1.13	1.53	1.68	28.8	32.4	26.1	29.4
7/8	22	1.40	1.54	2.08	2.29	39.0	43.8	35.4	39.7
1	26	1.82	2.00	2.71	2.98	50.7	56.9	46.0	51.6
1 1/8	29	2.31	2.54	3.44	3.78	63.6	71.5	57.7	64.9
1 1/4	32	2.85	3.14	4.24	4.67	78.2	87.9	70.9	79.7
1 3/8	35	3.45	3.80	5.13	5.65	94.1	106	85.4	96.1
1 1/2	38	4.10	4.50	6.10	6.70	111	125	101	113
1 5/8	42	4.80	5.27	7.14	7.84	130	146	118	132
1 3/4	45	5.56	6.12	8.27	9.11	150	169	136	153
1 7/8	48	6.38	7.02	9.49	10.4	171	192	155	174
2	51	7.26	7.98	10.8	11.9	193	217	175	197

*1 ton = 2000 lbs. 1 metric ton = 2204 lbs.

*To convert to kilonewtons (kN), multiply tons (nominal strength) by 8.896;
1 lb = 4.448 newtons (N).

Wire Rope
Compacted 19 Strand Classification

Compacted Strand • Rotation Resistant
Bright (Uncoated)

Nominal Diameter		Approximate Mass		Minimum Breaking Strength*			
				tons		metric tons	
inches	mm	lb/ft	kg/m	Standard	High Strength	Standard	High Strength
3/8	9.5	.31	.46	7.55	8.3	6.85	7.53
7/16	11.5	.40	.59	10.2	11.2	9.25	10.2
1/2	13	.54	.80	13.3	14.6	12.1	13.2
9/16	14.5	.69	1.03	16.8	18.5	15.2	16.8
5/8	16	.85	1.26	20.6	22.7	18.7	20.6
3/4	19	1.25	1.86	29.4	32.4	26.7	29.4
7/8	22	1.68	2.50	39.8	43.8	36.1	39.7
1	26	2.17	3.23	51.7	56.9	46.9	51.6
1 1/8	29	2.75	4.09	65.0	71.5	59.0	64.9
1 1/4	32	3.45	5.13	79.9	87.9	72.5	79.7
1 3/8	35	4.33	6.44	96.0	106.0	87.1	96.1
1 1/2	38	5.11	7.60	114.0	125.0	103.0	113.0

*1 ton = 2000 lbs. 1 metric ton = 2204 lbs.

*To convert to kilonewtons (kN), multiply tons (nominal strength) by 8.896;
1 lb = 4.448 newtons (N).

Wire Rope
Compacted Swaged 6 x 19 & 6 x 37

Compacted (Swaged) • Bright (Uncoated) • IWRC

Nominal Diameter		Approximate Mass		Minimum Breaking Strength*	
				EIPS	
inches	mm	lb/ft	kg/m	tons	metric tons
1/2	13.0	.55	.82	15.5	14.0
9/16	14.5	.70	1.04	19.6	17.8
5/8	16	.87	1.29	24.2	22.0
3/4	19	1.25	1.86	34.9	31.7
7/8	22	1.70	2.53	47.4	43.0
1	26	2.22	3.30	62.0	56.3
1 1/8	29	2.80	4.16	73.5	66.7
1 1/4	32	3.40	5.05	90.0	81.8
1 3/8	35	4.20	6.24	106.0	96.2
1 1/2	38	5.00	7.43	130.0	118.0

*1 ton = 2000 lbs. 1 metric ton = 2204 lbs.

*To convert to kilonewtons (kN), multiply tons (nominal strength) by 8.896;
 1 lb = 4.448 newtons (N).

Wire Rope Slings
1-Part Mechanical Splice

6 x 19 and 6 x 37 • EIPS • IWRC • Rated Capacity in Tons

	1 LEG		BASKET & 2 LEG BRIDLE			
Rope Diameter (inches)	Vertical	Choker	Vertical Basket or 2-Leg	60 degree	45 degree	30 degree
3/8	1.4	1.1	2.9	2.5	2.0	1.4
7/16	1.9	1.4	3.9	3.4	2.7	1.9
1/2	2.5	1.9	5.1	4.4	3.6	2.5
9/16	3.2	2.4	6.4	5.5	4.5	3.2
5/8	3.9	2.9	7.8	6.8	5.5	3.9
3/4	5.6	4.1	11	9.7	7.9	5.6
7/8	7.6	5.6	15	13	11	7.6
1	9.8	7.2	20	17	14	9.8
1 1/8	12	9.1	24	21	17	12
1 1/4	15	11	30	26	21	15
1 3/8	18	13	36	31	25	18
1 1/2	21	16	42	37	30	21
1 5/8	24	18	49	42	35	24
1 3/4	28	21	57	49	40	28
1 7/8	32	24	64	56	46	32
2	37	28	73	63	52	37
2 1/8	40	31	80	69	56	40
2 1/4	44	35	89	77	63	44
2 3/8	49	38	99	85	70	49
2 1/2	54	42	109	94	77	54
2 5/8	60	46	119	103	84	60
2 3/4	65	51	130	113	92	65

- Rated capacities based on design factor of 5.
- Rated capacities for basket hitches based on D/d ratio of 25.
- Rated capacities based on pin diameter no larger than natural eye width or less than the nominal sling diameter.
- Horizontal sling angles less than 30 degrees shall not be used.

Wire Rope Slings
1-Part Mechanical Splice

6 x 19 and 6 x 37 • EIPS • IWRC • Rated Capacity in Tons

Rope Diameter (inches)	3 LEG BRIDLE				4 LEG BRIDLE			
	Vertical	60 degree	45 degree	30 degree	Vertical	60 degree	45 degree	30 degree
3/8	4.3	3.7	3.0	2.2	5.7	5.0	4.1	2.9
7/16	5.8	5.0	4.1	2.9	7.8	6.7	5.5	3.9
1/2	7.6	6.6	5.4	3.8	10	8.8	7.1	5.1
9/16	9.6	8.3	6.8	4.8	13	11	9.0	6.4
5/8	12	10	8.3	5.9	16	14	11	7.8
3/4	17	15	12	8.4	22	19	16	11
7/8	23	20	16	11	30	26	21	15
1	29	26	21	15	39	34	28	20
1 1/8	36	31	26	18	48	42	34	24
1 1/4	44	38	31	22	59	51	42	30
1 3/8	53	46	38	27	71	62	50	36
1 1/2	63	55	45	32	84	73	60	42
1 5/8	73	63	52	37	98	85	69	49
1 3/4	85	74	60	42	113	98	80	57
1 7/8	97	84	68	48	129	112	91	64
2	110	95	78	55	147	127	104	73
2 1/8	119	103	84	60	159	138	112	80
2 1/4	133	116	94	67	178	154	126	89
2 3/8	148	128	105	74	197	171	139	99
2 1/2	163	141	115	82	217	188	154	109
2 5/8	179	155	126	89	238	206	168	119
2 3/4	195	169	138	97	260	225	184	130

- Rated capacities based on design factor of 5.
- Rated capacities for basket hitches based on D/d ratio of 25.
- Rated capacities based on pin diameter no larger than natural eye width or less than the nominal sling diameter.
- Horizontal sling angles less than 30 degrees shall not be used.

Wire Rope Slings
1-Part Mechanical Splice

6 x 19 and 6 x 37 • EIPS • IWRC • Rated Capacity in Tons
2 LEG CHOKER

Rope Diameter (inches)	Vertical	60 degree	45 degree	30 degree
3/8	2.1	1.8	1.5	1.1
7/16	2.9	2.5	2.0	1.4
1/2	3.7	3.2	2.6	1.9
9/16	4.7	4.1	3.3	2.4
5/8	5.8	5.0	4.1	2.9
3/4	8.2	7.1	5.8	4.1
7/8	11	9.7	7.9	5.6
1	14	13	10	7.2
1 1/8	18	16	13	9.1
1 1/4	22	19	16	11
1 3/8	27	23	19	13
1 1/2	32	28	23	16
1 5/8	37	32	26	18
1 3/4	43	37	30	21
1 7/8	49	42	34	24
2	55	48	39	28
2 1/8	62	54	44	31
2 1/4	69	60	49	35
2 3/8	77	66	54	38
2 1/2	85	73	60	42
2 5/8	93	80	66	46
2 3/4	101	88	71	51

- Rated capacities based on design factor of 5.
- Rated capacities for basket hitches based on D/d ratio of 25.
- Rated capacities based on pin diameter no larger than natural eye width or less than the nominal sling diameter.
- Horizontal sling angles less than 30 degrees shall not be used.

Alloy Chain Slings
Grade 80 Single Leg & 2 Leg

Working Load Limit, in Pounds • 4 to 1 Design Factor

Chain Size		SINGLE LEG		2 LEG		
		90 deg		60 deg	45 deg	30 deg
		vertical	choker			
inches	mm					
7/32	5.5	2100	1700	3600	3000	2100
9/32	7	3500	2800	6100	4900	3500
5/16	8	4500	3600	7800	6400	4500
3/8	10	7100	5700	12300	10000	7100
1/2	13	12000	9600	20800	17000	12000
5/8	16	18100	14500	31300	25600	18100
3/4	20	28300	22600	49000	40000	28300
7/8	22	34200	27400	59200	48400	34200
1	26	47700	38200	82600	67400	47700
1 1/4	32	72300	57800	125200	102200	72300

- Rated capacities for slings used in a choker hitch shall be a maximum of 80% of the rated capacities for single and multiple leg slings provided that the angle of choke is greater than 120 degrees.

- The horizontal angle is the angle formed between the inclined leg and the horizontal plane of the load.

Alloy Chain Slings
Grade 80 3 Leg & 4 Leg

Working Load Limit, in Pounds • 4 to 1 Design Factor			
	3 LEG & 4 LEG		
Chain Size	60 deg	45 deg	30 deg
inches mm			
9/32 7	9100	7400	5200
5/16 8	11700	9500	6800
3/8 10	18400	15100	10600
1/2 13	31200	25500	18000
5/8 16	47000	38400	27100
3/4 20	73500	60000	42400
7/8 22	88900	72500	51300
1 26	123900	101200	71500
1 1/4 32	187800	153400	108400

- Chain slings made with grades of steel chain other than Grades 80 and 100 alloy steel are not recommended for overhead lifting.

- Rating of multileg slings adjusted for angle of loading between the inclined leg and the horizontal plane of the load.

- Quadruple sling rating is same as triple sling because normal lifting practice may not distribute load uniformly on all four legs.

Eye Bolts
Forged Shoulder Nut and Machinery

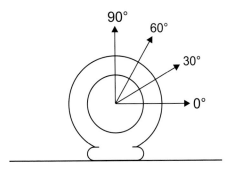

Forged Eye Bolts • Shouldered Type
Rated Capacities in Pounds

Nominal Size (inches)	90 degree	60 degree	30 degree	0 degree
1/4	400	75	NR	NR
5/16	680	210	NR	NR
3/8	1000	400	220	180
7/16	1380	530	330	260
1/2	1840	850	520	440
9/16	2370	1160	700	570
5/8	2940	1410	890	740
3/4	4340	2230	1310	1140
7/8	6000	2960	1910	1630
1	7880	3850	2630	2320
1 1/8	9920	4790	3840	3390
1 1/4	12600	6200	4125	3690
1 1/2	18260	9010	6040	5460
1 3/4	24700	12100	8250	7370
2	32500	15970	10910	9740

Shackles
Screw Pin and Bolt Type

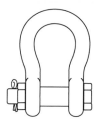

Screw Pin Anchor   **Bolt Type Anchor**

Forged with Alloy Pins

Nominal Shackle Size (inches)	Rated Capacity (lbs)
3/16	660
1/4	1,000
5/16	1,500
3/8	2,000
7/16	3,000
1/2	4,000
5/8	6,500
3/4	9,500
7/8	13,000
1	17,000
1 1/8	19,000
1 1/4	24,000
1 3/8	27,000
1 1/2	34,000
1 3/4	50,000
2	70,000
2 1/4	80,000
2 1/2	110,000

From the Crosby Group

NOTE: Use rated capacity marked on shackle if different than capacities listed in above chart.

Training and Certification

Crane Institute of America, Inc. is the leading provider of training and certification programs, services and products to users of lifting equipment in North America. Through our training and certification programs thousands of operators, inspectors, riggers and trainers have been trained and certified, all having a positive impact toward reducing accidents. Additionally, supervisors, managers and equipment owners have been provided with the training necessary to establish and improve safety programs on their job sites. Our training and certification programs are offered in two formats:

Scheduled Programs

These safety, inspection, management and Train-the-Trainer training programs are scheduled in major cities throughout the USA, including our headquarters near Orlando, Florida. Participants come from various industries throughout the world to these programs, which are conducted on rigging and a variety of cranes and equipment. Contact us for a detailed brochure or see our website for dates, locations and program descriptions.

On-Site Programs

These training programs are specifically developed for the customer's cranes and equipment and are conducted on-site at the customer's facility. Technical information is presented in a classroom setting followed by fieldwork where students are taught the basic and advanced techniques of actual crane operation, rigging, and inspection. Training can be conducted on all types of cranes and equipment for all levels of personnel. Contact us for a cost proposal.

Our safety, management, inspection and Train-the-Trainer programs are offered on the following types of cranes and equipment: mobile cranes, overhead and gantry cranes, tower cranes, portal cranes, aerial lifts, forklifts, basic and advanced rigging, scaffolding and fall protection.

Certification *is available at both our scheduled and on-site programs for operators, inspectors, riggers and trainers who meet the necessary requirements.*

Training and Certification

Crane Simulator Training and Certification

All of our Mobile Crane Safety training programs conducted at our headquarters near Orlando, Florida include hands-on training in our full motion crane simulator, which has a real operator's cab, actual controls, and a load moment indicator. Our simulator is so realistic you will think you are operating a real crane – and all this without the fear of having an accident.

Benefits of Simulator Training

- Provides an environment where students can learn the art of crane operation without the fear of damaging property or injuring people.
- Students can get immediate feedback regarding their performance from an unbiased instructor.
- Operating deficiencies can be targeted and corrected by repeating an operation.
- Crane operators can become more effective and efficient in the operation of an actual crane.
- Reduces the cost of taking an operational crane off-line for training purposes.

Training on our mobile crane simulator is only available to participants who attend our Mobile Crane Safety program in Orlando, Florida. Call 1-800-832-2726 to enroll.

Support Services

Legal and Investigative Support

Crane Institute of America employs highly trained experts in Accident Investigation and Reconstruction. As noted authorities in the safety and technical aspects of cranes and rigging, CIA personnel frequently provide expert testimony and litigation support, as well as interpretation of standards and regulations.

Site Safety Assessment

During our on-site visit we will assess your personnel, material handling practices, records and existing documentation for compliance with federal, state and industry requirements. Concluding our evaluation, an exit interview will be conducted followed by a written assessment and recommendations report.

A thorough safety analysis of your workplace helps prevent accidents before they happen.

Crane & Lifting Equipment Inspection, Load Test and Certification

An annual inspection conducted by an inspector from CIA followed by a written deficiency and recommendation report ensures OSHA compliance, exposes accident/liability potential and lists detailed corrective measures to ensure the equipment meets all requirements. ***Crane Institute of America, Inc. is accredited by the U.S. Department of Labor under Part 1919.***

In-House Training Programs

These custom training programs are developed by CIA for companies who train their crane and equipment operators and other personnel in-house. Experienced graphic designers use state-of-the-art software to produce full-color visual training aids which include viewgraphs, slides, videos, CD-ROMs and computerized presentation programs. Custom programs also include a student manual, instructor guide, exercises and job aids. Crane Institute will assist in the implementation of the program and train and certify in-house trainers.

Inspection Aids

- Inspection Certificate Decals
- Sheave Gauges
- Calipers (Stainless Steel)

- Checklists for Annual or Periodic Inspections:

Telescoping Boom Cranes	Vehicle-Mounted Aerial Lifts
Lattice Boom Cranes	Boom-Supported Aerial Lifts
Boom Trucks	Industrial Lift Trucks
Bridge & Gantry Cranes	Rough Terrain Lift Trucks
Jib Cranes & Hoists	Chain Slings
Monorail Crane Systems	Manual Chain Hoists
Tower Cranes	Lever Operated Hoists

- Checklists for Pre-Operational Inspections:

Telescoping Boom Cranes	Aerial Lifts
Lattice Boom Cranes	Industrial Lift Trucks
Overhead Cranes & Hoists	

- Checklists for Pre-Operational Procedures:
    Critical Lift Plan
    Personnel Platform Lift Plan

Safety Aids

- Rigging Safety Reference Cards
- Crane Safety Reference Cards
- Hoisting Personnel Safety Reference Cards
- Rigger's Capacity Cards
- Rigger's Capacity Wall Charts
- Hand Signal Cards (mobile and overhead)
- Hand Signal Charts (self-adhesive)
- Hand Signal Wall Charts (mobile and overhead)

Field Guides

Crane Institute of America has developed and published a series of high quality field guides on the following topics. Managers and trainers will find these guides ideal for toolbox talks and safety references.

- Crane Setup
- Working Cranes Near Power Lines
- Hoisting Personnel by Crane
- Pre-Operational Inspection
- Reductions in Rated Capacity
- How to Use Load Charts

Handbooks

Rigging

Companion volume to "Mobile Cranes"

This pocket-sized rigging handbook is well organized with easy to find information. Our down-to-earth language and use of graphics makes the data easy to understand and apply. This handbook contains the latest up-to-date information on wire rope, rigging hardware, slings and includes capacity tables and charts. This is a must-have for people who work with cranes and rigging.

Scaffolding Safety

Written to explain and illustrate OSHA's standard on scaffolding, this handbook is a must for all personnel, especially OSHA's required "competent person." Like all our books, the combination of graphics with text makes the information easy to understand and apply.

Fall Protection

This handbook is designed to discuss types, requirements and the application of standard fall protection systems. Using OSHA as a guide, it serves as a model for training personnel in effectively reducing the risks associated with fall hazards in the workplace.

Forklifts

Pocket-sized and user friendly, this comprehensive handbook is designed to provide operators, managers and safety professionals with the tools to identify and control lift truck related hazards.

Safety Training Videos

- Working Cranes Near Power Lines
- Pre-Operational Inspection for Telescoping Boom Cranes
- Pre-Operational Inspection for Lattice Boom Cranes
- How to Properly Interpret a Load Chart
- Hand Signal Communications
- Setup for Safety
- Overhead Crane Safety
- Advanced Rigging Tips
- Crane Work Rigging Techniques
- Rigging and Handling Structural Steel (Steel Erection)
- Rigging and Lifting with Mobile Construction Equipment
- Transport Trailer Safety - Transporting Equipment
- Hand Signals for Construction Equipment

ANSI/ASME Standards

We are authorized by ASME as a book dealer to distribute all ANSI/ASME standards and codes. Individual volumes cover the following:

- Jacks
- Overhead & Gantry Cranes
- Construction Tower Cranes
- Portal, Tower and Pedestal Cranes
- Mobile and Locomotive Cranes
- Derricks
- Base Mounted Drum Hoists
- Floating Cranes and Floating Derricks
- Slings
- Hooks
- Monorail and Underhung Cranes
- Handling Loads Suspended from Rotocraft
- Storage/Retrieval Machines and Associated Equipment
- Side Boom Tractors
- Overhead Hoists (Underhung)
- Overhead and Gantry Cranes
- Stacker Cranes
- Cableways
- Below-the-Hook Lifting Devices
- Manually Lever Operated Hoists
- Articulating Boom Cranes
- Personnel Lifting Systems
- Scrap and Material Handlers
- Low Lift and High Lift Trucks
- Rough Terrain Forklift Trucks

Crane Institute of America, Inc.
To order products call (800) 832-2726 or (407) 322-6800
Fax (407) 330-0660
www.craneinstitute.com

182